COMPASS

ALSO BY ALAN GURNEY

Below the Convergence
The Search for the White Continent

COMPASS

*A Story of
Exploration and Innovation*

ALAN GURNEY

W. W. NORTON & COMPANY
NEW YORK   LONDON

For information about permission to reproduce selections from this book,
write to Permissions, W. W. Norton & Company, Inc.,
500 Fifth Avenue, New York, NY 10110

Manufacturing by The Haddon Craftsmen, Inc.
Book design by Brooke Koven
Production manager: Amanda Morrison

Library of Congress Cataloging-in-Publication Data

Gurney, Alan.
Compass : a story of exploration and innovation /
Alan Gurney. —1st ed.
p. cm.
Includes bibliographical references and index.
ISBN 0-393-05073-4 (hardcover)
1. Compass—History. I. Title
VK577.G84 2004   2004004314
912'.028'4—dc22

ISBN 0-393-32713-2 pbk.

W. W. Norton & Company, Inc.
500 Fifth Avenue, New York, N.Y. 10110
www.wwnorton.com

W. W. Norton & Company Ltd.
Castle House, 75/76 Wells Street, London W1T 3QT

2 3 4 5 6 7 8 9 0

CONTENTS

Preface and Acknowledgments   9

Prologue   13

Chapter 1   Dead Reckoning   19

Chapter 2   Needle and Stone   31

Chapter 3   The Rose of the Winds   41

Chapter 4   Variation and Dip   55

Chapter 5   Edmond Halley, Polymath   69

Chapter 6   To Compass the Globe   77

Chapter 7   Halleyan Lines   87

Chapter 8   Dr. Gowin Knight and His Magnetic Machine   99

Chapter 9   Knight's Compass   109

Chapter 10   The Shocks of Tempestuous Seas   119

Chapter 11   Any Old Iron, Any Old Iron   135

Chapter 12   The Book of Bearings   149

Chapter 13   The Flinders Bar   161

Chapter 14   Soft Iron, Hard Iron   175

Chapter 15   "An Evil So Pregnant with Mischief"   187

Chapter 16   Deviation, the Hydra-Headed Monster   199

Chapter 17   The "Inextricable Entangled Web"   211

Chapter 18   Gray's Binnacle   225

Chapter 19   Thomson's Compass and Binnacle   235

Chapter 20   The Selling of a Compass   247

Chapter 21   A Question of Liquidity   261

Epilogue: From Needle to Spinning Top   273

Appendix: Deviation   277

Notes   281

Bibliography   297

Index   307

PREFACE

AND

ACKNOWLEDGMENTS

The genesis for this book sprang from two sources: one, the interest generated when researching my book on sailing exploration in the Antarctic, *The Race to the White Continent*, and the nineteenth-century concern with terrestrial magnetism; the other, listening, with a certain amount of incredulity, to the story of the high-tech yacht told in this book's prologue.

Thus I was led to the sudden realization, having spent many hours being guided by the compass on ocean races and coastal cruising, of my total lack of appreciation and understanding of all the developments which, over the centuries, have brought the magnetic compass to its present state of perfection. Well might John Hamilton have written in 1552 that "a skipper can not guide his ship to a good haven without direction of his compass."

Any writer on the magnetic compass and early navigation

is in deep debt to two men: Commander W. E. May R.N., founding member of the Royal Institute of Navigation and for many years Deputy Director of the National Maritime Museum; and Lieutenant-Commander D. W. Waters R.N., one-time Keeper and Head of the Department of Navigation at the same museum.

For specific help in gathering together the strands that make up this book I would like to thank the staff of the Public Record Office, Kew; the Hydrographic Office, Taunton; all the unfailingly helpful librarians at the Royal Geographical Society, The Royal Society, the National Maritime Museum, the British Library and, in particular, those at that wonderful institution, the London Library. I would also like to thank Dr. Toby Clark, Lady Joan Heath, Diana Harding, the Royal Yacht Squadron's Archivist, and the compass manufacturers Ritchie Navigation in the United States and Lilley & Gillie in England. And last, but most important, my wife, Carol, for her reading of the manuscript, her helpful suggestions, and her long-suffering patience in listening to my ramblings on the various individuals who figure in the following pages.

COMPASS

PROLOGUE

"The sailors, moreover, as they sail over the sea, when in cloudy weather they can no longer profit by the light, or when the world is wrapped in the darkness of the shades of night, and they are ignorant to what part of the horizon the prow is directed, place the needle over the magnet, which is whirled round in a circle, until, when the motion ceases; the point of it [the needle] looks to the North."

ALEXANDER NECKHAM, 1187

"To the east . . . is the boundless ocean, where the sun and sky blend their colours, and the passing ships sail only by means of the south-pointing needle. This has to be watched closely by day and night, for life or death depend on the slightest fraction of error."

CHAO JU-KUA, 1225

"To guide them and help them regain the shore they have but one preservation, a mariner's compass, for with this they can see which way to set their course after the winds have risen."

OLAUS MAGNUS, 1555

"Who was the first inventor of this instrument miraculous, and embued as it were with life, can hardly be found."

WILLIAM BARLOWE, 1597

"The compass is the soul of the ship."

VICTOR HUGO, 1866

"Curiously enough, until comparatively recently, shipbuilders and owners did not bestow on the compass the amount of consideration which it undoubtedly merits. It is *pre-eminently* the instrument upon which the safety of the vessel depends, and justly ranks *first* in importance. It would be easier to dispense with the chronometer, or even the Sextant, than with this invaluable guide."

CAPTAIN S. T. S. LECKY, 1908

"There is no more fascinating subject than the history of man's going down to the sea in ships and then out across it. It is the story of maritime peoples everywhere."

ANDREW SHARP, 1956

On a sunny spring morning of 1998, a new and very expensive yacht set off on her sailing trials. This was a vessel brimming with the latest in high-tech wizardry and it would have taken a very clever diviner to find the few pounds of natural materials, wood, wool, and cotton, among the tons of petroleum derivative foams, fibers, and resins that had gone into the construction of her hull, deck, accommodation, sails, mast, and rigging.

The navigation station, with its screens, keyboards, dials, flickering digital numbers, and softly glowing lights, was a monument to defense industries, microchips, orbiting satellites, and James Bond movies. No inconvenient and cumbersome charts, tide tables, parallel rules, plotters, dividers, or pencils cluttered this altar to modernity. The navigator, sitting before an array of screens, could happily tap at the ranks of keyboards to find the yacht's position on the chart glowing on one screen, then set the course and punch the command into the self-steering mechanism. On other screens he could note the yacht's speed, the wind's speed and direction, the state of the tide and depth of water under the keel; study a weather map; monitor approaching craft on the radar.

On deck the human helmsman (when required) steered his course by a fluxgate compass. This compass—and its vast array of attendant gadgetry—was powered by electricity from an equally vast array of batteries.

On the first night at sea, a moonless, starless night, cloud-covered and black as pitch, all the electronic gadgetry failed due to a defective switchboard. Screens went blank. Digital read-outs faded. The fluxgate compass, losing its life-giving electrical pulse, metaphorically rolled-up its eyes and died.

The owner had put all his trust into high-tech navigation, forgetting (or perhaps not even being aware) that Bran Ferren, a computer expert, had memorably defined technology as "stuff that doesn't work yet." Not even an old-fashioned, hand-bearing magnetic compass disfigured the bulkhead.

The helmsman, as the yacht sliced through the waves in thoroughbred style, suddenly realized that his only sense of direction came from the wind. The reassuring glow from the compass and wind-direction indicators had vanished. Lighthouses and buoys, blinking their signals, were all far below the horizon. No stars or moon glittered in the heavens to act as celestial beacons. His only reference point had been reduced to the wind blowing on his right cheek. With all on board praying that the wind stayed directionally steady, the yacht was put about. With the wind now blowing on the helmsman's left cheek and with an anxious torch shining on a scrap of fabric tied and fluttering on a shroud (a very low-tech wind indicator), the blinkered thoroughbred, under reduced sail, sailed back on what was hoped a reciprocal course. Early that morning the clouds cleared and the Pole Star, with a welcome glitter, showed them north, followed by the sun lifting itself above the eastern horizon. Later that day, with the aid of binoculars and a crew member familiar with the waters, the yacht returned to the marina.

The loss of the compass proved a most traumatic experience. Sailing blind, by wind direction alone, is the stuff of dinghy sailing. Offshore, out of sight of land, on a pitch-black night with no directional guides save the wind, is another experience. An experience that took those twentieth-century yachtsmen back a thousand years. Back into a time when men navigated across the seas before the marine compass had been invented.

This is the story of man and the marine magnetic compass; the story of an instrument which guided countless sailors across seas and oceans into the unknown. The story of an instrument so precious to northern seamen of the sixteenth century that any man found tampering with the compass or its magnetizing lodestone had his hand, by law, pinned to the mast with a dagger, the even more painful result being a split palm as the offender, to gain freedom, dragged his hand down against the blade.

Chapter 1

DEAD RECKONING

The cry of "Breakers ahead!" was the first warning that the navigating officers had made a dreadful mistake in their dead reckoning, or, in the mordant Spanish equivalent, their *navegación de fantasia*.

Minutes later, in the howling dark of an autumn night, four ships of a Royal Navy fleet were mastless hulks being pounded to pieces between the hammer blows of the Atlantic breakers and the anvil of the Scilly Islands' granite reefs. Some two thousand men and officers from HMS *Association*, *Eagle*, *Firebrand*, and *Romney* died on that night of October 23, 1707. One more vessel, HMS *St. George*, struck hard but surged clear and survived.

Among the drowned was the fleet's portly, florid-faced commander, the fifty-seven-year-old Admiral Sir Cloudesley Shovell. His body was found washed ashore in a sandy bay some seven miles from the wreck. Legend has it that a local woman, thirty years later, made a death-bed confession that she had found a waterlogged Sir Cloudesley unconscious on the shore, and then helped him into eternity for the sake of the diamond and emerald rings on his fingers.

The death by drowning of two thousand men and the loss of ships is still considered the worst shipwreck disaster ever suffered by the Royal Navy. It led the British government, by a 1714 Act of Parliament, to create the Board of Longitude and its financial prizes for anyone discovering a "practicable and useful" method of determining a ship's longitude at sea. The end result was threefold: the publishing of accurate astronomical tables for calculating the lunar distance method of finding longitude; John Hadley's reflecting quadrant for the accurate measurement of lunar distances; and John Harrison's famous marine chronometer.

Ironically enough, the shipwreck disaster was not so much a matter of longitude but more a matter of latitude, inaccurate charts, an unknown current, and shoddy compasses.

THE GATEWAY OF the English Channel, from the Scilly Islands in the north to Ushant in the south, makes an opening of 100 nautical miles. But Ushant, for Shovell's grand fleet, had to be avoided. All seamen avoided Ushant, with its reefs and strong tides. Ushant also happened to be French, and Britain was at war with France. Shovell, having broken the French fleet in the Mediterranean, was returning home for the winter with a fleet of twenty-one ships. On September 30, after sailing from Spain, the fleet headed north into gales and overcast weather.

By October 21, they were on soundings of between 90 and

140 fathoms. Shovell now called a meeting of the sailing masters to hear their combined opinion as to the fleet's position. At the mouth of the English Channel, south of the Scilly Islands and due west of Ushant, was their verdict. Only Captain Sir William Jumper of HMS *Lennox* disagreed. He claimed that the Scilly Islands were much closer and would be raised in a few hours. He went unheeded. The *Lennox* and two other ships were then ordered to Falmouth for convoy duties. One of them, HMS *Phoenix*, ground across some of the Scilly Islands' rocky outliers but won clear to safety with her pumps working overtime.

At four o'clock in the afternoon of October 22, Shovell ordered his remaining vessels to heave-to and take soundings. They lay there for two hours, rolling in the building swells, satisfied that they were clear of all dangers. Then they squared away before the favorable gale to run up Channel. Two hours later, breakers were sighted, foaming around rocks. Guns were fired, the recognized danger signal. But it was all too late.

THIS CATASTROPHE FOR the Navy—and the nation—can only be understood in the context of the times and its tools of navigation for finding a dead-reckoning position.

A dead-reckoning position (or D.R., in navigators' language) is a position arrived at by estimates of distance traveled and course steered, the log giving the distance traveled and the compass the direction steered. Added to this the navigator has to make allowances for currents and, if a sailing vessel, the leeway. When approaching land or sailing in shoal waters he uses another aid, a depth sounder.

Shovell's fleet carried that most ancient of navigating tools, the lead line, for measuring the depth of water. Like all good inventions it has the grace of simplicity. The lead weight, usually weighing 7 pounds for a 25-fathom line and 14 pounds for a 100-fathom line, is cup-shaped at its base. This hollow is filled with tallow (known as arming the lead) so that when the lead strikes bottom, particles of the seabed stick to the tallow: sand, mud, stones, shells, etc. These particles provided vital information for coastal sailing and are still to be found on British Admiralty charts. Sprinkled among the numerals on the chart, giving the depth of water, can be found cryptic and mysterious letters: *S* for sand; *M* for mud; *Si* for silt; *St* for stones; *Sh* for shells; *Oz* for ooze; and, giving a hint of more pellucid and warmer waters, *Co* for coral. Hydrographers can even become garrulous, in a nautical text-messaging style, with mixed types of seabed: *fS.P.bkSh.G.Ck*, for instance, meaning fine sand, pebbles, broken shells, gravel, and chalk.

This information, perhaps, sounds rather academic for the sailor who piously hopes to sail across the surface of the sea rather than sink to the bottom. But in the days when sailors navigated along coasts and across seas in a mood of fearful optimism, every scrap of information, even to the color of mud, could be of life-saving importance.

The 25-fathom lead line is still marked in a traditional manner: two strips of leather at 2 fathoms; three strips at 3 fathoms; one square of leather with a hole in it at 10 fathoms; white duck at 5 and 15 fathoms; red bunting at 7 and 17 fathoms; blue serge at 13 fathoms; a piece of cord with two knots in it at 20 fathoms. By this method, in Shovell's day, a seaman could tell the depth at night by the feel of the markers. A longer lead line, the deep-sea lead line, was marked in multiples of 10 fathoms with a cord having a number of knots equal to the multiples. The lead line, in short, enabled a vessel to find her way, even at night, by feeling her way through the underwater contours of known sandbanks, shelves, and channels. Radar shows the modern mariner the shape of the coastline; the old-fashioned lead line showed the early mariner not only the shape of the bottom but also its composition.

The earliest English manuscript rutter (or book of sailing directions) dates from the early fifteenth century and among its information, which must date back into much earlier times, it describes the type of bottom to be expected after coming on soundings—100 fathoms—when sailing on a passage from Gibralter to England. The detail as to the type of bottom is astonishing. Off Penmarch Point in Brittany, 60 fathoms would produce "sandy ooze and black fishey stones" and 50 fathoms just "black ooze." Off Belle Isle, also in Brittany, 60 fathoms would produce "small dial sand"—the type of fine sand used in hour-glasses. Off Portland Bill in the English Channel, at 24 fathoms, would be found "fair white sand with red shells therein."

An early illustration of the lead line appeared in 1584 on the title page of what was to become a most famous and innovative atlas of sea charts. Published by Lucas Wagenaer, a Dutch pilot, it was soon translated into English as the *Mariner's Mirrour*. It became such a nautical *vade mecum* that for the next two centuries any atlas of charts, for English seamen, was known as a "waggoner."

The title page of Wagenaer's *Spieghel der Zeevaerdt* shows two seamen surrounded by sixteenth-century navigating instruments: quadrants, sea astrolabes, sandglasses, cross staves, celestial and terrestrial globes, dividers, magnetic compasses. One of the seamen, dressed in a long cloak and pointed hat, looks more like a bearded Merlin engaged in cabbalistic incantations surrounded by his instruments of magic, summoning monsters from the deep—also illustrated—than a seaman demonstrating the use of the lead line. This made a very appropriate title-page illustration, for an air of mystery, of magic, of wizardry, surrounded the art of navigation across the world's seas and oceans.

Not shown in the illustration is an English invention for measuring a vessel's speed and thus her distance run: the log. The first mention of this instrument appears in William Bourne's *A Regiment for the Sea,* published in 1574; the first illustration appears in Samuel de Champlain's *Les Voyages de la Nouvelle France Occidentale*, published in 1632. In his description Champlain wrote that he had seen it used by "several good English navigators" and the results were much superior to the "reckonings ordinarily made." Prior to the English log the "reckonings" had been made by throwing overboard a chip of wood and then timing, usually by counting, how long it took to pass along a known distance marked on the bulwarks. The log, by comparison, was a sophisticated instrument. It consisted of a wooden segment of a circle of about nine-inch radius with the curved bottom edge weighted with lead. This simple device was attached by three small lines, like a crow's foot, to a long line. When thrown overboard the segment floated upright. The long line had knots spaced at a set distance apart (commonly 48 feet) and was wound around a reel. To measure the vessel's speed the log was dropped overboard at the stern, the reel being held by a seaman, bracing himself for the jerk as the log hit the water, and a half-minute sandglass turned. The number of knots that unreeled before the sand ran out of the glass gave the ship's speed in nautical miles per hour. A ship's speed is still given in knots.[1]

The third item to find the dead reckoning was the magnetic compass. Today's marine magnetic compass is a liquid compass with its card floating in a liquid (usually a mix of water and alcohol), the liquid dampening the card's movement. The compasses used in Shovell's day were dry compasses—the compass card, with its magnetized needle underneath, pivoting on a vertical pin set in a glass-covered wood or brass bowl. As sailing ships have a horrid tendency to roll and pitch in a seaway, the compass bowl was hung in gimbals (double pivoted rings) to keep the compass card as horizontal as possible.

And it was toward these compasses that Captain Sir William Jumper, the dissenter when it came to the dead-reckoning position of Shovell's fleet, had pointed his accusing finger. Jumper happened to be a widely respected officer whose opinion could not be taken lightly. The fleet's compasses were called in to be inspected and the report made terrifying reading. Of 136 wood-bowl and 9 brass-bowl compasses, only 3 were found effective. Of 370 wood-bowl compasses in store at Portsmouth, only 70 were found to be in good working

[1] The alert and mathematically minded reader will have noticed something wrong. The nautical mile is 6,080 feet. For a half-minute glass the knots should be spaced at 51 feet. But seamen, practical men, shied away from impractical numbers. Forty-eight feet happens to be 8 fathoms. If fractions were needed for the distance between knots, these would be in eighths, a fathom. Log-book speed columns were always headed *K* (knots) and *F* (fathoms). The other advantage of the 48-foot spacing was that the ship's speed was overestimated, thus putting the vessel ahead of her actual position, something of an advantage when approaching land.

condition. Compasses were found to be defective when supplied to the Navy yards. On some compasses the gimbals were
so weak that they broke when issued to a ship. One compass
maker claimed that his rival's gimbals were "no thicker than
an old Groat." Wooden bowls were cracked; compass needles
were so weakly magnetized that after a few months they were
useless; compass cards, supported underneath by the compass
needle (aligned north and south), sagged at the east and west
sides. One of the reasons for this sorry state of affairs was the
custom of keeping spare compasses in the boatswain's storeroom. This was invariably damp and close to the powder
room, where the niter in the gunpowder speeded up the rusting process. It was suggested, to no avail, that compasses be
placed in seasoned wood boxes and the boxes stored in the
ship's bread room. But, as a vague nod to the critics, the Admiralty did decide to buy only brass-bowl compasses.[2]

Gunfire and thunderstorms could also demagnetize needles. The report suggested that lodestones for remagnetizing
the needles—or, as it was charmingly known, "refreshing the
compass"—be supplied to ships expected to be away from a
Navy yard for a year or more. Navy yards, in turn, should buy
lodestones and employ a suitable person to remagnetize and
repair compasses.

But the best compass in the world is still of small use if the
charts are incorrect. And on most charts of Shovell's day the
Scilly Islands were shown some miles north of their true position. Throw in a then-unknown current, the Rennell Current, that runs north from Ushant to the Scilly Islands after

[2]Wood-bowl compasses were being used by the United States Navy
until 1830.

westerly gales in the Bay of Biscay, plus a strange effect called magnetic variation, and it is not so surprising that Shovell ran into disaster.

Magnetic compasses point to the magnetic north pole, not the geographic north pole. The angle made by the difference between these two poles is known as the magnetic variation. The angle varies east or west of the line to the geographic pole, the line of longitude, and this angle is known as east or west variation. For navigators it is of huge importance and must be allowed for when setting a course. To complicate matters even further, magnetic variation is not a constant. It varies across the globe. It also—yet another great bane to seamen—changes over the years: the secular change. Magnetic variation at London in 1580 was 11°15' East. By 1773, it had swept through 32 degrees to 21°09' West. By 1850, it had increased to 22°24' West. A hundred years later it had decreased to 9°07' West. It is still decreasing today.

The navigator ignores the magnetic variation at his peril. The astronomer and mathematician Edmund Halley called attention to this in 1701, a few years before the Shovell disaster. Magnetic variation in the Channel, he pointed out, had changed from east to west in 1657 and now stood at about 7½° West. In other words, an older seaman, used to little or no variation, steering due east by compass and thinking this his true course, would, in fact, be steering north of east.

Shovell's navigating officers, showing a cavalier disregard for good navigational practice, failed to make any allowance for magnetic variation. The combination of the Scilly Islands being south of their charted position, the Rennell Current's sweeping them north, and the foolish navigator neglecting the variation, spelled death and disaster for Shovell, his men, and his fleet.

TWO HUNDRED AND sixty years after the drowning of those two thousand men from Shovell's fleet, the tanker *Torrey Canyon*, carrying 120,000 tons of crude oil, drove at 16 knots, in broad daylight, perfect visibility, and a calm sea, onto one of the granite Seven Stone rocks lying off the Scillies. The massive oil spill could be smelt far into Devon. No lives were lost except those of sea birds, fish, and crustaceans.

The *Torrey Canyon* was owned by the Barracuda Tanker Corporation, whose corporate office was a file in a drawer in a filing cabinet in a building on Gorham Road, Hamilton, Bermuda. The owners of the corporation were New York lawyers. Barracuda Tanker Corporation had leased the tanker to Union Oil of California. She was on a single-voyage charter to British Petroleum, which was 49 percent owned by the British government. The tanker had been built in the United States and lengthened in Japan to increase her capacity. She was manned by an Italian crew and sailed under the Liberian flag. This macedoine of an enterprise was to lead to her being the second largest vessel ever to be lost at sea.

Her navigating equipment consisted of radar, loran, radio-direction finder, depth-sounder, and sextants. For most of the voyage from Kuwait the tanker had been steered by an automatic pilot operating off a gyrocompass. A gyrocompass points to true north, thus eliminating all the headaches of allowing for magnetic variation. The automatic steering had twice given trouble during the long voyage.

The Scillies happen to be well signposted with lighthouses

and lightships. But when it was obvious that his ship was heading toward disaster—radar and compass bearings were giving ominous plots on the chart, and fishing boats were sending out warning signals—the captain shouted an order to alter course. Nothing happened, for the steering mechanism was not on manual. By the time the change had been made it was all too late. According to the Board of Investigation the tanker's life was now numbered in seconds: "The bow of the *Torrey Canyon* began to swing left. According to the course recorder graph, she reached a heading of 350 degrees when at 8:50 the vessel struck Pollard Rock and came to a sudden stop hard and fast aground." The remains of Sir Cloudesely Shovell lying in Westminster Abbey—under what the *Dictionary of National Biography* has called "an elaborate monument in very questionable taste"—might have given a twitch. The Scillies could claim another shipwreck to go in the record book.

Navigation has become more sophisticated since Shovell's day when compass needles were remagnetized by a lodestone. But the path from lodestone to global positioning systems has been a tortuous one. One marked by wrecks and sailors' bones.

Chapter 2

NEEDLE AND STONE

Lodestone, loadstone, magnetite, oxide of iron, Fe_3O_4—all describe a dull, gray-black ore that can be found as an outcrop of stone above ground. This unattractive rock has remarkable characteristics. Not only will it attract iron but it can magnetize iron. It also exhibits polarity. A long, slim sliver of lodestone, if suspended at its middle by a thread, will align itself north and south. This same lodestone sliver, if stroked against a metal needle, will transfer to the needle the same characteristics. The stroker, in other words, will have produced a compass needle.

Some specimens of lodestone, like some people, are possessed with an extra dollop of what might be called lithic sex appeal in the shape of super magnetism. Magnet Creek in Arkansas is famous for rocks pumped full of this magnetic libido.

The fifth-century Saint Augustine of Hippo, a man with enough libido to father an illegitimate son, tells of being astonished to see a lodestone pick up a metal ring, this ring picking up another ring, and yet another ring, until a chain of rings hung from the lodestone, all held by an unseen force. The uncanny power of this force was also described to Augustine by a fellow prelate. According to him, scraps of metal, spread across a silver plate, were moved around by the power of a lodestone held underneath the plate.

Roger Bacon, the brilliant thirteenth-century English philosopher and scientist who gained such a reputation as an alchemist and magician that he was jailed by the Church, experimented with lodestones. One end of the stone, he found, might draw iron to it, but the other end would "make it flee like a lamb from a wolf," a delicious metaphor for observing that like poles repel and unlike poles attract. He also noted that the attractive power worked through water. An iron needle, thrust through a straw and floated in a bowl of water, would dive underwater when the lodestone was held under the bowl, and rise up when the lodestone was held above.

But the lodestone, to some medieval minds, held far more important qualities than those demonstrated by Bacon. If placed on the pillow of a wife suspected of adultery it would lead to a sleep-talking confession. A small pulverized amount taken in sweetened water would melt away fat. A similar amount taken in plain water would restore youth. This most marvelous panacea cured gout and headaches, prevented baldness, drew the poison from wounds, relieved pain, and eased childbirth. It made men and women gracious and elegant speakers. Burglars found it helped their thieving, for a small amount burned in a house made the householder think the house collapsing, and then flee. A lodestone carried in one's pocket would cause melancholy. When mixed with nettle juice and snake fat, it would make a person mad and send the unfortunate soul wandering from family, home, and country. The odor of garlic and onion would demagnetize a lodestone, leading to the suggestion that sailors be prevented from eating these pungent, if healthy, vegetables so as not to play havoc with the compass needle and the ship's navigation.

ALEXANDER NECKHAM, AN English scholar-monk, was the first to record a metal needle being magnetized by a lodestone and then used as a marine compass. Alexander was born in 1157, on the same day as the future King Richard I of England, Richard the Lion-Heart. Alexander's mother suckled both infants: Richard at her right breast and Alexander at the left. Twenty-three years later, Neckham was teaching at the University of Paris while his foster brother was refining the finer points of medieval warfare and mayhem in Aquitaine.

Neckham the scholar wrote two treatises typical of their time. *De Naturis Rerum* was a treasure chest of legends and folk tales: the Man in the Moon; the last song of the dying swan; the development of the goose from a barnacle; the sharp-eyed lynx that can see through nine walls; the squirrel who crosses a river on a plank of wood holding up his tail as a sail. *De Utensilibus* dealt with articles in everyday use.

In *De Naturis Rerum* he described how "the sailors, moreover, as they sail over the sea, when in cloudy weather they can no longer profit by the light of the sun, or when the world is wrapped in the darkness of the night, and they are ignorant to what part of the horizon the prow is directed, place the needle over the magnet, which is whirled round in a circle, until, when the motion ceases, the point of it [the needle] looks to the North."

In *De Utensilibus* the compass needle is described as mounted on a dart, "thus making known to the sailors the route which they should hold while the Little Bear is concealed from them by the vicissitudes of the atmosphere." Neckham's dart has caused much head-scratching among academics: Was the dart a vertical dart, meaning a pivoted compass needle? Or did he mean the needle was stuck through a dart or straw, thus making the needle float in a bowl of water?

The various and learned arguments by the academics, pro or con, have produced far more words than those written by Neckham. As Neckham lived in an age of monkish and scholarly dispute, these arguments give a certain ironic twist to the question. What is known is that Neckham had a profound distaste for the sea, thought it dangerous and that voyages should only be undertaken in dire necessity. His crossings of the English Channel were the probable cause of this virulent dislike.

A few decades after Neckham, two Dominican friars described an iron needle being thrust through a straw to make a cross, and then floated in a bowl of water. The lodestone, after being brought to the side of the bowl, was moved around the bowl, faster and faster, until the needle was whirling swiftly. At this point the lodestone was quickly drawn away. And then, wrote the friars: "The needle turns its point towards the Stella Maris. From that position it does not move."

In 1218, Jacques de Vitry, a French bishop sailing to his church in Acre, wrote that an iron needle, "after it has made contact with the magnet stone, always turns towards the North Star, which stands motionless while the rest revolve, being as it were the axis of the firmament. It [the needle] is therefore a necessity for those travelling by sea."

Even poets wrote about the magnetized needle and its uncanny ability to point to the Pole Star. The French troubadour Guyot de Provins, writing at the beginning of the thirteenth century, satirically wished the Pope to be as constant as the Pole Star, which sailors

> watch whenever they can, for by it they keep course. . . .
> By the virtue of the magnet-stone they practise an art which cannot lie. Taking this ugly dark stone, to which iron will attach itself . . . they find the right point on which they touch with a needle. Then they lay the needle in a straw and simply place it in water, where the straw makes it float. Its point then turns exactly to the Star. There is never any doubt about it, it will never deceive. When the sea is dark and misty, so that neither Star nor Moon can be seen, they put a light beside the needle, and then they know their way. Its point is towards the Star, so that the sailor knows how to steer. It is an art that never fails.

The art must have taken time to evolve. The finding that a sliver of lodestone pointed north; that an artificial lodestone could be made by rubbing an iron needle with the natural lodestone; that the needle would turn if floated in water; that a certain end would always point to the Pole Star; that the iron needle had to be rubbed with a specific end of the lodestone—all this must have been a long process of evolution, and one certainly not arrived at in a single leap of imagination or experimentation.

THE CHINESE, centuries earlier than any mention of the fact in Western literature, were aware of the directional properties given to a metal needle when touched by a lodestone. The encyclopedist Shen Kua, in his *Mêng Chhi Pi Than*, writing in the late eleventh century, noted that: "A geomancer rubs the point of a needle with the loadstone to make it point south" and then went on to say that the needle could be floated on the surface of the water but was rather unsteady. The needle could also be balanced on a finger nail or rim of a cup. But the best method was to suspend it by a single thread of silk fastened by a blob of wax at the needle's center. Hung in a windless place it would point to the south.

But balancing a magnetized needle on a finger nail and holding up a needle in a windless place is party-trick material, stuff that, perhaps, might be ideal for a geomancer demonstrating his powers of divination, but that a seaman would find comic to try at sea.

Other Chinese texts of the eleventh century mention magnetized fish-shaped, wafer-thin iron leaves floating in shallow water-filled bowls, the fish head pointing south. A century later (at roughly the same time as the mariner's compass was being recorded by Western monks, friars, and poets), Chinese magicians were making floating wooden fish fitted with a hidden lodestone. The fish head pointed south. Another construction, and perhaps even more delightful, was a small wooden turtle, again fitted with a hidden lodestone, which balanced and rotated on a sharpened bamboo pin. The turtle's head pointed north and its tail south. These enchanting instruments were not used at sea, but were part of the stock in trade of magicians and geomancers.

The first definite mention of a marine compass in Chinese literature occurs in the twelfth-century *Phing-Chou Kho Than* (Phingchow Table-Talk) written by Chu Yü, whose father had been a high official in the port of Canton. One chapter describes commercial voyages between Canton and Sumatra: "The ship's pilots are acquainted with the configuration of the coasts; at night they steer by the stars, and in the day-time by the sun. In dark weather they look at the south-pointing needle." Chu Yü then continues with a description of the Chinese equivalent of the European lead line: "They also use a line a hundred feet long with a hook at the end, which they let down to take samples of mud from the sea-bottom; by its appearance and smell they can determine their whereabouts."

By the end of the thirteenth century, Wu Tzu-Mu, in a description of sailing from Hangzhou into the East China Sea—"a vast expanse of sea without a shore, strong and very dangerous, the abode of mysterious dragons and marvellous serpents"—wrote graphically of the dangers to sailors among its reefs and low-lying islands. He then ends with lip-smacking relish on the finality of a pilot's mistake: "It depends entirely on the compass, and if there is a small error you will be buried in the belly of a shark."

The compass that these unfortunates relied upon was the simple, free-floating needle as described by Neckham. It was not until the late sixteenth century that the Chinese adopted the European dry-pivoted compass. Japanese pirates had noted the superiority of the Dutch and Portuguese compasses and started using them in preference to their free-floating needles.

As one sixteenth-century commentator wrote: "Recently the people of Chiangsu, Chekiang, Fukien and Kuangtung have all suffered from the attacks of Japanese pirates. Japanese ships always use at the stern the dry-pivoted compass to fix their course; and our people, having captured this, imitated it so that the method became common in Chiangsu."

IN 1269 (SOME few years earlier than Wu Tzu-Mu's writing on the dangers of sailing into the East China Sea), Petrus Peregrinus (Peter the Pilgrim), born Pierre de Maricourt, a soldier in the French army besieging the southern Italian fortress town of Lucera, wrote a long letter to a friend in Picardy. The letter had wide influence and over the years many copies came to be made, one of them being a particularly beautiful copy in Oxford's Bodleian Library. The letter concerned magnets and compasses.

One of the compasses described consisted of a carefully cut and shaped lodestone fitted into a circular sealed box with a pointer on the top lid. The box was then floated in a slightly larger bowl of water. This contrivance was designed to show astronomers the true meridian without using the sun. The second compass was more sophisticated. This was a pivoted, magnetized needle placed in a glass-lidded box. The edge of the lid was graduated into degrees and was provided with sights for taking bearings of stars.

Peregrinus was also wrestling with another idea. The lodestone, he thought, carried within itself the likeness of the heavens with its stars and planets; and so always pointed toward the celestial pole. A correctly cut and mounted lodestone, in the shape of a sphere, should revolve in absolute syn-

chronicity with the earth: in other words, a perpetual clock.
His friend Roger Bacon thought it would be worth a king's
treasury. But, alas for Peregrinus, the revolving lodestone
refused to move.

A BOWL OF water holding a straw transfixed by a magnetic
needle is hardly a satisfactory instrument for use at sea. The
makeshift cross soon bumps against the bowl's side and with
the vessel pitching and rolling the water soon slops out of the
bowl. The magic of the "needle and stone," as it was called,
touched with a hint of the occult as the pilot bent over the
bowl, looking as if he was stirring a pot as he whirled the
lodestone around the bowl, was only to locate north and
check that the wind direction had not changed. Vessels in
Neckham's day, when all the celestial beacons had vanished,
steered by the wind. The floating compass was nothing more
than a very valuable aid.

Chapter 3

THE ROSE OF THE WINDS

Within a quarter mile of the Parthenon, overlooking the labyrinthine, dusty, narrow lanes and stairways of the Plaka district in Athens, rears the octagonal Tower of the Winds. Built about 100 B.C., its eight battered faces are carved with the eight personified winds of the Classical world: Boreas, Kaikos, Apeliotes, Euros, Notos, Lips, Zephyros, Skiron. Boreas, the blustery north wind, blows on a conch shell and wears a cloak with its folds flapping in the wind. Kaikos, an icy wind from the northeast, empties a shield full of hailstones. Apeliotes, the mild east wind, holds corn and fruits. Euros, the southeast wind, has an arm wrapped in a mantle and threatens a storm. Notos, the south wind and bringer of rain, empties an urn of water. Lips, the southwest wind, holds the *aphlaston* (a ship's stern ornament), promising a rapid voyage. Zephyros, the mild west wind, strews a lapful of flowers. Skiron, the dry northwest wind, holds an urn filled with glowing charcoal.

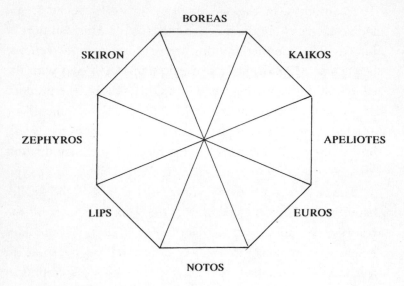

The eight winds of the Greeks as personified on the octagonal Tower of the Winds in Athens.

Even the most urbanized Athenian of the Hellenic world, reeling out from an evening's conversation at a symposium, replete with food and wine, full of pleasant memories of delicious dalliance with slave boys, dancing girls, and the more polished *hetairia* (the Athenian equivalent of an expensive prostitute), would have felt these winds on his cheeks, and recognized them by their strength, humidity, temperature, and direction—but not, perhaps, to the same extent and accuracy as a sailor.

THE ANCIENT MEDITERRANEAN sailors' sailing season was a short one, a season compressed into the seven months between

April and October. During the winter months the galleys were hauled out and merchant ships laid-up alongside quays. Those months, in the age of Plato, Aristotle, Socrates, and Demosthenes, were the quiet months for the Athenian port of Piraeus. The summer months, by contrast, were a scene of hectic activity, pungent with exotic smells and loud with the babble of different tongues. On the quayside lay the produce of the lands surrounding the Mediterranean and the Black Sea: flax, hemp, papyrus, dates, nuts, incense, cheese, figs, wine, honey, olive oil, hides, ivory, timber and pitch, and, most important of all, grain from southern Russia, Sicily, and Egypt.

The freighters for these cargoes varied in size, the smaller being coasting vessels such as the one sunk by pirates off the Cyprus coast about 300 B.C. Her replica, *Kyrenia II*, built after extensive underwater archaeological research, is only some 46 feet long. The original was carrying four hundred amphorae of wine, olive oil, and ballast. She had about four crew members, who tended a rectangular sail set on one mast. She was steered by two steering blades set port and starboard of the stern, a method that continued to be used in the Mediterranean until the fifteenth century A.D.

Over the centuries the freighters became larger. Imperial Rome, a city with over a million people, gobbled some 400,000 tons of grain annually. A third of this came from Egypt, hauled by a fleet of large grain carriers (vessels equivalent to today's oil tankers in comparative size, and also, like today, providing the populace with an essential for their daily living). Other vessels were wine carriers fitted with clay tanks. The tanks weighed a ton empty and held 800 gallons of wine when full.

These freighters could only make effective progress with a

following, or quartering, wind. A head wind stopped them as effectively as a cliff. In the eastern Mediterranean and Black Sea the prevailing summer winds are northerly—the Greeks called them Etesian, or "annual," winds—which made the bringing of grain to Athens from the Black Sea a relatively easy voyage, but that from Egypt a long and tedious one.

The grain voyages to feed Rome were equally bedeviled by the prevailing northwest winds of the central Mediterranean. From Ostia—Rome's port—to Alexandria would take about two weeks of fair sailing, but the return voyage, two months of hard and circuitous sailing. Seven centuries after the Tower of Winds had been built, such sailing was hateful to the Arab general who had conquered Egypt. "Trust it little, fear it much," he warned, "man at sea is an insect on a splinter, now engulfed, now scared to death."

Over the centuries the Mediterranean seamen, before and after sailing from their harbors, had put their trust, like Homer's heroes, into a libation to the gods sprinkled across deck and sea. Plus, hedging their bets, trust in sailing directions—*periplus*—handed down over the generations.

Agathermos, a Greek geographer, wrote that "from Paphos to Alexandria is 3,800 stadia with Boreas." In other words, the northerly wind would take the vessel from Paphos, on the south coast of Cyprus, the 300 miles to Alexandria.

As the center of influence and political gravity moved from Greece to Italy, the winds changed their names and the sailing directions became more sophisticated and comprehensive. Pliny the Elder, the first century A.D. naturalist, wrote that "from Carpathos it is 50 miles with Africus to Rhodes," Africus being the southerly wind. During the Crusades, Bishop Jacques de Vitry's vessel would have followed sailing directions on entering Acre's harbor to stay clear of the rocks off St. Andrew's church and then, when the "Constable's house is in direct line with the Tower of Flies (the tall minaret of a mosque) you can make straight for the harbor."

A SAILOR AT SEA, out of sight of land, sits at the center of a circle whose circumference is the horizon. This circle is quite small. Standing on deck with his eyes 9 feet above the sea, he is at the center of a watery, circular world some 7 miles in

The eight winds of the Mediterranean as seen by Italian sailors of the thirteenth century.

diameter. From the masthead, say 75 feet above the sea, his world has increased to 20 miles in diameter.

It was across this circle that blew the Mediterranean winds of the sailor: the eight of the Tower of Winds (although the more intellectual toyed with a twelve-wind system), which soon became, by dividing, a sixteen-wind system. For the thirteenth-century Italian sailor the winds blowing across his circular world were named after sunrise in the east (Levante), sunset in the west (Ponente), the sun at noon in the south (Mezzodi or Ostro). The north wind came from over the northern mountains and was named Tramontana. Dividing the four quadrants of the horizon made by these four cardinal winds were other winds, the half-winds: Greco (northeast), Sirocco (southeast), Libeccio (southwest), Maestro (north-west). These were the eight principal winds. By combination, sixteen winds could be designated: such as the musical *Sirocco ver levante poco*, southeast and a little east.

One of the reasons for this multiplicity of winds was the improvement in the sailing qualities of ships due to a number of technical developments. A rudder hung at the stern instead of the two steering blades; the introduction of the bowline for hauling tight the leading edge of a squaresail when the yard was braced around for winds other than those coming from astern or on the quarter; the adding of one or two more masts, one of them being the lateen-rigged mizzen mast. All these improvements enabled square-rigged ships to sail closer to the wind and increase their arc of sailing. Using the analogy of a clock face, a merchant ship of Aristotle's time, with a wind blowing from the north or twelve o'clock, had a sailing arc restricted to the hours between four o'clock and eight o'clock. Later square-rigged ships, with the new technological improvements of rig and rudder, had increased that arc of sailing from two o'clock to ten o'clock. Yards could be braced around, bowlines hauled tight, and the attempt made to sail upwind. It was an evolution which ended with one of the world's most beautiful artifacts: the short-lived, graceful, fast, wind-driven nineteenth-century tea and wool clippers, vessels doomed to be swept from the world's oceans by another technology based on iron and steam.

ON NOVEMBER 27, 1095, at Clermont-Ferrand in central France, Pope Urban II exhorted Western Christendom to make an armed pilgrimage to the Holy Land to wrest back from the Turks the Holy City of Jerusalem. Four years later, during July 1099, the Holy City fell to the Christian knights and foot soldiers of the First Crusade. Jerusalem's fall was due, in large part, to the June arrival at Jaffa of a Genoese fleet carrying food, timber, and skilled labor, the timber and labor for the building of massive siege towers, catapults, and *ballistae* (giant crossbows).

One of the results of Jerusalem's capture was the setting up of the Latin Kingdom of Jerusalem. For the next two hundred years, until its fall in 1291, this feudal state required a lifeline and support system; the main artery for this support was the Mediterranean Sea and the ships from the republican maritime city-states of Pisa, Genoa, Venice, and Amalfi. All these city-states grew fat off the Crusades; and their merchants performed the dangerous juggling act of prudently reinforcing their Christian compatriots in Jerusalem while imprudently hazarding their immortal souls by trading with the Muslims.

All this vast coming and going of fleets martial and commercial led to a rapid improvement in sailing directions and charts. The first mention of a chart's use appears in 1270, during a Crusade led by France's King Louis IX with a fleet of thirty-nine vessels chartered from owners in Genoa and Marseille. The ships' pilots, in the calm after a storm, had to calm a worried Louis by showing him a chart and pointing out to the king that they were off Sardinia's Cagliari Bay.

The oldest existing chart, the Carta Pisana, has been dated close to 1275. It extends from the Black Sea to southern England. The coasts of the Black Sea and Mediterranean are extraordinarily accurate; less so the Atlantic coastline. The chart is drawn on a sheepskin with the neck on the right (an exception to the usual sheepskin chart, where the neck is on the left). This is a working chart designed and drawn by someone, probably Genoese, familiar with mathematics and the needs of the sailor.

To the modern sailor's eye its appearance is bewildering. The Carta Pisana has no lines of longitude or latitude, although it does have a scale. Its most distinctive features are two great circles, one centered in the Aegean Sea and the other near Sardinia. Sixteen lines radiate out from the centers. Where they meet the circumference they are reflected back like rays, giving the chart a slightly hallucinatory effect. With a dollop of imagination the resulting circle, with its sixteen points like pointed petals, looks like a stylized flower head. Here is the rose of the wind, or wind rose.

Later charts were simplified and the web of lines made easier to identify with color. The coloring tended to be standardized. The eight principal winds were black, the eight half-winds green, the sixteen quarter-winds red. Miniature wind roses became highly decorative, with their eight, sixteen, or even thirty-two points being colored. The wind rose now looked like a star.

The rhumb lines, or loxodromes, running out from the wind rose were used by the navigator, with the aid of a rule and pair of dividers, to find his course from port to port. The Italian-born explorer Amerigo Vespucci (whose Latinized name, Americus, was given to the New World by the German cartographer Martin Waldseemüller) paid 130 gold ducats for such a loxodromic chart.

These Mediterranean charts would be accompanied by *portolans* (pilot books, or sailing directions). The oldest surviving portolan dates back to 1296 and gives sailing directions for the circuit of the Mediterranean in a clockwise direction from Cape St. Vincent on the Iberian Peninsula to Safi in Morocco. The directions read just like modern sailing directions: dangers are noted, safe anchorages are suggested, courses and sailing distances given. Such charts and sailing directions, based on the information gathered by the use of the magnetic compass—the charts and directions are too accurate for the information to have been collected without this instrument—boosted the economy of Europe.

The merchant ships of Venice, Genoa, Pisa, and Amalfi, following the example of the Greeks and Romans, had also avoided sailing during the winter months. One fleet would sail for the Levant at Easter and return in September. Another fleet, the winter fleet, would sail from Venice in August, winter overseas, and return to Venice in May. A Genoa fleet would sail for Egypt and the Levant in September, winter overseas, and return to Genoa in June. Pisa had a rule that any vessel arriving in harbor after November 1 could not put to sea until March 1.

All this had changed by the fourteenth century. The compass had brought the confidence, and the ability, to sail in overcast weather. Fleets were now making two round voyages a year. Pisa had ships leaving in the middle of winter. "Thus," writes Frederic Lane, a world authority on all things related to waterborne Venice, "the traditional 'closing of the sea in winter,' which had persisted in the Mediterranean during several thousand years, was shattered by the compass."

THE COMPASS DESCRIBED by Neckham, a magnetized needle stuck through a straw and floated in a bowl of water, would have been far too crude an instrument for collecting the information provided by the sheepskin charts and sailing directions. For this information the magnetized needle was balanced on a vertical pin in a dry bowl. But another refinement was added: an unknown genius made a circular card similar to the wind rose drawn on charts, and then placed it over the needle. The sailor could now steer a specified course, a compass course, and take bearings of objects ashore using the points of what was now a compass card. The Portuguese still call the compass card a *rosa dos ventos*: a wind rose.

It is more than probable that this innovation was made by an Italian.

Amalfi is a small fishing village perched in a ravine on the north shore of the Gulf of Salerno. It overlooks an extravagantly blue sea that attracts tourists and celebrities. In the ninth century, Amalfi was a powerful maritime city-state rivaling Pisa, Venice, and Genoa. It had a maritime law, the *Tabula de Amalpha*, which became universal throughout the Mediterranean. But Amalfi's short reign as a trading city-state began its decline with the sacking of the city by Pisa, followed by an earthquake and storm in 1343 which destroyed the harbor. The final blow came in 1348 with the arrival of that great killer, the plague known as the Black Death.

To offset this bruising history, Amalfitanes boast that one of them, in 1302, invented the magnetic compass. And the town celebrated this fact six hundred years later in 1902 with a bronze statue of a bearded and hooded figure peering into a compass held in his hands. This is the statue of Flavio Gioia, the supposed inventor of the magnetic compass. Amalfi also boasts a Piazza Flavio Gioia and a hotel La Bussola (The Compass).

A request to the British government to help pay for these celebrations was given short shrift by the Admiralty Hydrographer. He was not alone in this. For Timoteo Bertelli, an Italian historian who had spent three decades delving into dusty archives, now revealed to Amalfi that Flavio Gioia was nothing but a myth, a myth created over the centuries by the fog which surrounds and distorts manuscripts written centuries after the supposed facts. No matter, the celebrations went ahead, and today the mythical Flavio Gioia faces the Mediterranean and gazes at his compass: a maritime version of King Arthur with his sword.

NORTHERN EUROPEAN SAILORS never adopted the Italian names for the rhumbs of the wind. Their circular horizon was divided with the Germanic north, south, east, and west. These were soon combined to give greater accuracy. Alfred the Great, the ninth-century Anglo-Saxon king, used the terms northeast, southeast, northwest, and southwest, giving an eightfold system.

Centuries later, Geoffrey Chaucer, in his 1390 "Treatise on the Astrolabe," wrote that English sailors divided the horizon into thirty-two parts. The names for these, using the simpler "southeast by south" rather than the more musical Italian *sirocco ver levante poco*, led the Portuguese and Spanish to adopt the northern ways. The French, with a Mediterranean and Atlantic seaboard, continued to use both systems until the early eighteenth century. And the Italians hung on to their system until the nineteenth century.

Wind rose from Italian sailing chart (circa 1492) with initials of the winds, a stylized fleur-de-lis marking north and a cross marking east.

But the Mediterranean, having created the chart wind roses, still continues its ghostly influence. The fleur-de-lis is the standard mark showing the north point of the compass card. Wind roses on the Mediterranean charts were marked this way, the general theory being that the *T* for Tramontana, marking north, became embellished by some unknown draughtsman into a fleur-de-lis. The other mark of the wind rose, a cross or some other embellishment to mark the east point, lasted into the nineteenth century, as did map decorations showing the personified Italian winds—cherubs and old men with puffed-out cheeks.

"A SQUARE RIGGED Vessel when close hauled (i.e. as close to the Wind as she can possibly lie), can approach no nearer to it than six Points; to the perfect understanding of which, the Young Sea Officer must make himself thoroughly acquainted with the Mariner's Compass, which should be diligently got by art, that he may refer to it in his Memory on all Occasions." So wrote Darcy Lever in his 1808 *The Young Sea Offi-*

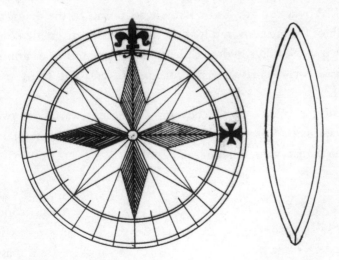

Compass card marked with 32 points, a fleur-de-lis marking north and a cross marking east. The oval magnetic needle is shown on the right. From *Breve Compendio de la Sphera de la arte Navegar* by Martin Cortes (1551).

cer's Sheet Anchor. The art was "diligently got" by long and tedious memorizing—akin to the memorizing by children of the multiplication tables—in an exercise known as "boxing the compass," where the young seaman had to recite the thirty-two points of the compass clockwise from north, and then counterclockwise. The 360-degree compass card ended that particular form of mental torture.

Just as we still retain vestiges of a reptilian brain—the R-complex—so does a modern chart retain a vestigial rose of the winds. This is the circular compass rose, marked in degrees rather than compass points. On many charts there are two superimposed compass roses. One points to geographic north, "true north" in navigator-speak. A smaller compass rose sits within the larger. This smaller compass rose, its north point marked by a fleur-de-lis or arrowhead, points to the magnetic north of the area covered by the chart. On this magnetic compass rose, kindly hydrographers have printed the annual increase or decrease of the magnetic variation from true north.

For the matter of the changing magnetic variation, over the centuries, was yet another problem sent to tax the brains of mathematicians, scholars, and seamen.

Chapter 4

VARIATION AND DIP

A whiff of brimstone and sulphur surrounds Dr. Dee. The alliteration of his name alone conjures up visions of an urbane, suave, and demonic villain pitting his wits against James Bond. It's a wonder that Ian Fleming never used it. (But perhaps, in a way, he did.)[1]

John Dee, born in 1527, was an Elizabethan scholar, mathematician, astrologer, alchemist, geographer, magus, and spy. In a superstitious age, Dee made the astrological calculations to

---

[1]Queen Elizabeth I, in an age fond of nicknames, called three men her "Eyes": Sir Christopher Hatton, known as "Lids," signed his letters to her with two triangles with dots inside; the Earl of Leicester, with two circles with dots inside; Dee, with two circles and the lengthened top stroke of the number 7 across their tops: 007. The number seven for Dee was a cabbalistic and lucky number. Elizabeth called Dee "my noble Intelligencer."

select the most propitious day for Queen Elizabeth I's coronation. Elizabeth, a woman who had the measure of most men, was somewhat in awe of her horoscope caster. And Dee, acting as a wise counselor, was often called to her court: to confer on her health; to suggest means of preventing possible mischief when an image of her body, with a pin stuck through the breast, was found in Lincoln's Inn Fields; to calm a worried court on the alarming appearance of a comet.

As a mathematician he gave free lectures at the College of Rheims in France, lectures so popular, so crowded, that many were forced to listen and peer through the windows. As a scholar he suggested a national library, anticipating Thomas Bodley's Oxford library by fifty years and the British Museum Library by two hundred years.

He had gained his reputation as a magus by creating astounding stage effects for the theater, consulting what he called his "magic mirror" and communing, using a rather sinister young man as a medium, with a crystal sphere (now in the British Museum) to call forth evil spirits and angels. Dee, somewhat disingenuously, denied being a "companion of the hellhounds, and a caller, and conjuror of wicked and damned spirits." But his reputation as a sorcerer led a London mob to storm his house at Mortlake, where they then destroyed many of his possessions, including globes from his friend Mercator and a five-foot-diameter quadrant once owned by Richard Chancellor, the chief pilot for the Muscovy Company. The mob then turned to ruining a large part of his four-thousand-volume library.

They also stole "a magnet-stone, commonly called a loadstone, of great virtue, which was sold out of my library for 5 shillings, and for it afterward, yea piece-meal divided, was more than £20 given." The lodestone reflected Dee's consuming interest in geographical discovery and navigation. As an advisor on navigational problems to the Muscovy Company, he taught and became friends with a host of Elizabethan scholars, seamen, and adventurers: Sir Walter Raleigh, Sir Humphrey Gilbert, Sir John Hawkins, Richard Hakluyt, Richard Chancellor, John Davis, Stephen Borough, William Borough, Christopher Hall, and Martin Frobisher. He devised the circumpolar chart used by the Muscovy Company's navigators in northern waters and a compass card divided into degrees as well as the usual points.

He also coached Frobisher and Hall, prior to their 1576 voyage in search of a northwest passage, in the use of some unfamiliar navigational instruments. Frobisher's portrait shows him as the epitome of the spade-bearded Elizabethan seaman dear to the hearts of small boys. On his left hip rests a fearsome-looking sword and his right hand grips an even more fearsome-looking horse pistol. On a nearby table stands a globe of the world. Here is a Yorkshireman who would, measured by today's standards, be quickly brought up before a magistrate on a most comprehensive charge sheet, a charge sheet that would have included piracy. But no charge could have been brought against him for poor seamanship.

One of the instruments used by Frobisher and Hall was a "great instrument of brass named Compassum Meridianum" for measuring the angle between the geographic and magnetic poles and twenty other compasses "of divers sorts." The meridional compass was to be put to good use. On June 12, 1576, anchored in the Thames off Gravesend Castle, Hall measured the compass variation to be 11½ degrees to the east.

Today, in Hatfield House, some 20 miles north of London,

can be seen the results of the magnetic variation measurements taken on that northern voyage. It's a sheepskin chart
prepared by William Boroughs, an experienced Arctic pilot
and agent for the Muscovy Company. The coasts of Ireland
and England are colored blue, France red, Scotland, Norway,
and Greenland green. The chart has the usual wind roses and
loxodromic lines. Also shown are small arrows indicating the
compass variation from true north. It is the first such chart
showing magnetic variation.

THIS WANDERING POLE was a matter of great perplexity to
seamen, geographers, astrologers, and astronomers. Some
blamed it on the quality of the lodestone used in "touching"
the compass needle; some thought that the compass needle,
having been magnetized, pointed to the northern area where
the lodestone had been dug from the earth; some speculated
that magnetic mountains lay in the northern regions, mountains of such prodigious magnetism that ships sailing nearby
would have the nails drawn from their deck and planking;
some thought it an error caused by poor navigation; some
blamed currents setting ships off course; and some derided it,
believing no such thing existed.

By the middle of the fifteenth century the Portuguese—
those pioneers of oceanic sailing—were colonizing the Azores
and sailing their small caravels far down the African coast. And
their pilots noted the compass needle pointing away from true
north: the northeasting or northwesting, as they called it.

At the same time German craftsmen were making pocket
sundials—a sort of early portable watch—that allowed for
variation. These lovely artifacts were fitted with a small com-

pass. Engraved on the box was a variation mark; the compass needle was set to this mark and thus the small sundial would be set in its correct north–south alignment. It only worked accurately, of course, if the pocket sundial was used in the same part of Europe as the variation marked on the box. But then this was an age closer to the rhythms of nature than today; most people could tell the time by the length of their shadows, and some clocks had only an hour hand.

This cunning idea of offsetting the compass to allow for variation was also used by the marine compass makers of Flanders.

IN THE SUMMER of 1556, Stephen Borough was sailing along the Russian coastline in an absurdly small pinnace named the *Searchthrift*. As chief pilot of the Muscovy Company, he was hoping to find trading lands via a northeast passage. South of Novaya Zemlya the *Searchthrift* was stopped by ice. But some days before he had landed at the mouth of the Pechora River and measured the magnetic variation, "which was $3\frac{1}{2}°$ from the north to the west."

Twenty-four years later, his brother William was also taking measurements for magnetic variation, but this time in the less exotic area of London's Limehouse district. William Borough, shortly to be nominated Comptroller of the Navy, used a special compass of his own design and recorded a variation of $11\frac{1}{4}°$ to the east.

The London of 1580 was a stinking, brawling, vital, and noisy city sucking into itself not only the country's produce but also its vagrants, beggars, and unskilled labor. It also absorbed skilled artisans and craftsmen: men with agile fingers and brains, ready and able to manufacture the new breed of astronomical and navigating instruments. Such a man was Robert Norman, who had spent twenty years at sea and then decided to set up shop to produce charts and compasses. He called himself, in a mildly self-deprecating fashion, a mere "unlearned Mechanician." The variation compass used by William Borough for his measurements had been made by Norman.

It was Norman's practice when making a compass to ensure that the needle and card were perfectly balanced, the card horizontal, before magnetizing the needle. But he found that after the needle had been magnetized the north end of the compass card would dip—and stay dipped. The phenome-

Compass card of the sixteenth century. The broken double line indicates the lozenge-shaped needle offset from north to allow for variation.

non intrigued him. But nothing in his reading of all the experts offered an answer. So Norman just added a small piece of wire at the south end as a counterbalance. One commission came to him which required a long needle. He cut and polished it, balanced it perfectly, and then magnetized it. As usual, it dipped down. This time he pared away at the north end to restore the balance, but cut away too much. "Stroken with choler" at having ruined a compass needle and wasted both time and money, he decided to investigate.

After some thought (and advice), he made a 6-inch needle, finely balanced, through which ran a brass axle. He mounted this needle with the axle horizontal and the needle vertical. Behind the needle he placed a card marked off in degrees. His instrument was then turned into the magnetic meridian and the needle magnetized. The needle dipped in a most satisfactory manner: 71°50', to be precise. To counter critics who thought the lodestone, by some mysterious means, had merely added weight to the needle's end, Norman had weighed the needle before and after making his experiment. He recorded no change in weight.

William Borough had been an eager follower of Norman's experiments and persuaded him to make his findings public. In 1581, Norman's *Newe Attractive*, dedicated to William Borough, was published. Norman dealt with variation as well as dip, and bound into the small volume was *A Discourse of the Variation* by Borough.

MAGNETIC VARIATION, at the time Norman published his book, had been known for at least 150 years, possibly longer. But the discovery of the New World, the fishing voyages to

Iceland and Newfoundland, the Portuguese voyages to the East Indies, the circumnavigations by Magellan and Drake, had all added to the gathering of data on the strange and wandering ways of the compass needle.

The Portuguese and Spanish, regarding the French and English as poachers in their colonial seas, kept any knowledge of magnetic variation very close to their chests. But by 1542 enough was known that a line of no variation, now known as the agonic line, where the magnetic needle points to true north, ran north and south through the Atlantic. Two years later, Spain's Pilot-Major, Sebastian Cabot, published his famous planisphere showing the agonic line close to the Azores.

Here, it was thought, was a true meridian from which, using compass variation, a ship's longitude could be calculated. Based on very little evidence, and a large amount of faith, French, Spanish, and Portuguese pilots argued that the change of variation from the agonic line was gradual and equal. Measure your vessel's compass variation and the distance away from the agonic line could be found. One French pilot calculated that the compass needle would alter one degree for every 22½ leagues. This, for the converted, was the answer to the navigator's Holy Grail: finding a ship's longitude at sea.

Both Norman and Borough gave this method short shrift. Variation, they wrote, was dismayingly irregular, particularly so in the high northern latitudes. Its measurement was not helped by compass makers who had copied the German portable sundial makers' method of allowing for variation. These compass makers had made their correction for variation by offsetting the needle (hidden by the compass card) from the fleur-de-lis marking north. The offset varied as to where the compass had been made. Norman listed the types of compasses available for the unwary navigator. The Levant compass, made in Sicily, Genoa, or Venice, with no offset to the needle; the others, from as far apart as Danzig, Flanders, England, Seville, Lisbon, Rochelle, Bordeaux, and Rouen, having their needles offset east of north by anything from a full compass point (11¼ degrees) to half a compass point. Compasses like this had been in use for over a century. And charts had been made using such compasses. Norman warned his readers of using a compass from one area with a chart from another. In other words, a Levant compass should be used with a Mediterranean-made chart. An English compass with such a chart, or a Levant compass with an English chart, could only lead to what Borough dryly called a "confused mingle-mangle" and Norman, being more direct, to the sailors' "great peril."

Norman, in order to clear the minds of confused mariners as to what magnetic variation meant, made a simple diagram to illustrate the problem. His idea is still used on modern charts.

Norman and Borough were experienced sailors. But a total neophyte, a man who cheerfully admitted that he loathed the sea, was soon to make significant improvements and give sage advice on the mariners' compass. For William Barlowe, a man who was to spend his life as an eminent clergyman, secretly desired a different life. Barlowe was a seaman manqué, a man fascinated by the sea and its problems of safe navigation. In 1597, he published *The Navigators Supply*, a volume of his improvements on navigators' instruments, followed in 1616 by the *Magneticall Advertisments* on the lodestone and compass needles.

In both books, as a good clergyman, he anathematized the inaccurate compasses of his day with their crude workmanship: cards unequally divided; unbalanced cards balanced by blobs of wax; warped cards, their sides drooping; rusting-wire compass needles; needles in unequal oblong shapes; cracked glass covers. The list is enough to make any reader wonder that sailors survived even the shortest of voyages.

Barlowe, having studied this sorry material that could lead a sailor to his death, set about changing the odds. He designed navigating instruments, polar charts, and compasses. He explained the difference between iron and steel needles; improved the needle's shape; made an easily removable card so the needle could be easily remagnetized; gave instructions as to the best method of remagnetizing the needle by stroking it with the lodestone three or four times from the needle's center to the ends, using the north end of the lodestone for the needle's north end, and the south for the south. He also designed an azimuth compass for measuring the variation which happened to be an improvement on the instrument designed by Norman and Borough; it was a compass with sights and a verge ring marked in degrees, the first such compass, and was to be used by grateful seamen for over two hundred years.

But Barlowe the churchman (he was the son of a bishop and his four sisters all married bishops) surfaces with the words from Psalm 107 under an illustration for an instrument he called the Traveler's Jewel. "They that goe down to the Sea in Ships, and employ their labour in the great waters, They see the Workes of the Lord, and his wonders in the depe."

IN 1573 THE thirty-three-year-old Dr. William Gilbert, Fellow of the Royal College of Physicians, settled down in London. Born in Colchester, educated at Cambridge, and given a worldly polish in Europe, Gilbert was a bachelor of means and equipped with an inquiring mind. Some of the means (five thousand pounds' worth) went on buying books, manuscripts, globes, charts, lodestones, and instruments, all to be used during years of experiments and in putting together a treatise on terrestrial magnetism: Norman's *Newe Attractive* having set him on this unusual course.

In 1600, Gilbert published his magnum opus, *De Magnete*. Written in Latin—the language of science—it brought praise from Galileo ("I think him worthy of the greatest praise for the many new and true observations which he has made") and a half-century later some lines from John Dryden ("Gilbert shall live, till loadstones cease to draw, / or British fleets the boundless ocean awe"). Gilbert, for his part, gave the English language the words *electricity*, *electrical force*, *electrical attraction*, and *magnetic pole*.

Gilbert claimed, rather alarmingly, that the earth was a giant magnet. He had arrived at this hypothesis after carrying out hundreds of experiments. One was making artificial magnets by hammering iron rods laid down north-and-south in the magnetic meridian. Another was to instruct a lapidary to make a small-scale model of the earth from lodestone, complete with poles and marked with meridians and lines of latitude. Gilbert called his models terrellas. But the less sophisticated preferred to call them by the more homely "earthkin." Across these earthkins Gilbert moved small compasses and dipping needles, and then measured the variation and dip.

These lodestone earthkins were used by Gilbert to demonstrate to an enchanted Elizabeth I and her courtiers the earth's magnetic power, and his earthkins soon became a seventeenth-

Making an artificial lodestone. Woodcut from W. Gilbert's *De Magnete*. It shows a blacksmith facing north (Septentrio) hammering a red-hot iron bar laid in the north-south direction, thus making an artificial lodestone or compass needle.

century fad. Ben Jonson, in his comedy *The Magnetic Lady*, had a heroine, Lady Loadstone, and other characters named Needle, Compass, and Ironside. Terrellas became fashion accessories for mistresses. John Evelyn, in a diary entry, wrote of a "pretty terrella, described with all the circles and showing all the magnetic variations." Pepys, in his diary entry for November 2, 1663, wrote: "This day I received a letter from Mr. Barlow with a terrella which I had hoped he had sent to me. But to my trouble I find it is a present from him to my Lord Sandwich; but I will make a little use of it first, and then give it to him."

Gilbert, following hard on the heels of Norman, and working with William Barlowe, had designed and built an improved, sea-going dipping needle. With this, he claimed, a pilot could find his latitude in snow, fog, mist, or rain merely by using an engraved brass disk that contained information on magnetic dip calculated from experiments with a large terrella.

But the finding of one's latitude and longitude by magnetic dip and variation was based on quicksand. This was an age when scientific hypotheses were argued from small amounts of evidence. And the great tragedy of science (as later pointed out by Thomas Huxley) is the slaying of a beautiful hypothesis by an ugly fact.

The ugly fact emerged in 1633. Henry Gellibrand, an English mathematician (dismissed by the *Dictionary of National Biography* as "a plodding industrious mathematician, without a spark of genius"), found that the variation in London had decreased from Borough's 1580 measurement of $11\frac{1}{4}°$ East to 4° East. Other measurements also showed a perplexing decrease in variation. What was once thought to be static was proving to have an alarming and skittish behavior. Gellibrand published his findings and concluded that the variation had a built in variation: the secular change.[2]

Gellibrand's findings, nevertheless, failed to slay the beauti-

---

[2]For those interested in such matters, the secular change has a small seasonal fluctuation known as the *annual* change. To make matters even more complicated, there is also a *diurnal* change where the variation is most easterly at about 8 a.m. and most westerly at 1 p.m. In winter the diurnal change is smaller than in summer. And *local attraction*, or *local magnetic anomaly*, is used to explain abnormal variation due to large chunks of ferrous rocks.

ful idea of finding longitude by variation. On the contrary: Henry Bond, a teacher of navigation, thought it a positive help. Having studied the decrease of variation he foretold that the variation in London, by 1657, would be zero; and then slowly increase to the west. He gained great kudos by being right: and then moved on to draw up a theoretical table of magnetic elements. This table, with a compass for measuring variation and a dipping needle, would result in a mariner's being able to find his longitude. And all this could be found in his volume *The Longitude Found*, published in 1676—followed, two years later, by Peter Blackbarrow's deflating *The Longitude Not Found*.

But the idea of finding one's position at sea with a compass and dipping needle refused to die. In 1721, William Whiston published *The Longitude and Latitude discovered by the Inclinatory or Dipping Needle*. Whiston was no stranger to the milder forms of lunacy. After the wrecking of Shovell's fleet he, with another deluded man, had suggested a crack-brained scheme of anchored ships strung across the Atlantic. These ships would fire rockets into the sky timed to explode at a certain height. Any mariner seeking his position would take a compass bearing and note the time between the flash and the sound of the explosion, thus obtaining his distance and bearing from the ship. Whiston's absurd scheme reflects the man. Learned, admired by his contemporaries for his honesty and simplicity, he had the flawed intellect ("ill-balanced intellect," according to the *Dictionary of National Biography*) of the naive.

If there is a touching naivete about Whiston, the same cannot be said of Edmond Halley. Ten years Whiston's senior, Halley was to spend a lifetime, and sail thousands of miles, in studying the effects of the earth's magnetism on the compass.

Chapter 5

EDMOND HALLEY,
POLYMATH

In the autumn of 1691 the thirty-five-year-old Edmond Halley was walking along the bottom of the English Channel. For warmth, he was dressed in a double layer of woolen underclothing, a watertight suit of oiled leather, and a girdle of lead weights, all topped by a large helmet trailing two flexible pipes (made from layers of animal gut and wire) that vanished into a nearby diving bell.

Halley, peering through the murky waters at a depth of sixty feet, was searching for ivory and gold from a sunken ship of the Royal African Company. The diving suit and diving bell were his invention. Why Halley, an astronomer, was breasting the underwater currents off the Sussex coast is a cause for wonder. But Halley, over his long lifetime, was destined to wear many hats and engage himself in many activities. In an age of polymaths, he was the polymath supreme.

The slightly built Halley, of the humorous mouth and quizzical eyes, was no stranger to the sea. In 1689 he had surveyed the approaches to the Thames estuary and presented his chart to the Royal Society, along with a claim that he had corrected the many mistakes of previous charts.

The breadth of his interests can be seen in his communications to the Royal Society during the year he was diving on the wreck: papers on the transits of Venus and Mercury across the sun's face to determine the sun's distance from the earth; on the physical mechanism of evaporation; on measuring the thickness of gold upon gilt wire; on Pliny's book of natural history; historical detective work, using astronomy and tides, to locate the place and time of Julius Caesar's landing on England's coast; the speed of a bird's wing in flight; the measurement of wind and water forces; the refraction of light; the height to which bullets could be shot and fountains made to squirt; and a report on his diving activities where he had three men working for nearly two hours at a depth of ten fathoms.

A Londoner, born in 1656, he had lived through apocalyptic times of plague, fire, and war. The Plague of 1665, which killed seventy thousand Londoners and left the survivors with dreadful memories of the cry "Bring out your dead!" and the rumbling of the corpse-filled carts carrying their grisly cargoes to mass graves. Followed in 1666 by the Great Fire of London, which raged for four days, reduced the City between the Tower and the Temple to smouldering ashes, and made homeless a quarter of a million people: an event that caused rejoicing in Paris and Amsterdam. The Lord Mayor of London, with one of those heroic misjudgments of which history is so full, when woken to see the flames, stumped back off to bed grousing, "Pish! A woman might piss it out." A year later, wealthy merchants were packing their belongings and fleeing with their families on hearing the news that the Dutch fleet had burned and sunk laid-up warships in Chatham and, the ultimate insult, carried away the *Royal Charles* in triumph.

Halley, at sixteen, demonstrated both nautical and scientific precocity by measuring and tabulating the magnetic variation at London and other ports. A year later he entered as a commoner in Queen's College, Oxford. A contemporary, Anthony Wood, wrote of Halley: "He not only excelled in every branch of classical learning, but he was particularly taken notice of for the extraordinary advances he made at the same time in the Mathematicks. In so much, that he seems not only to have acquired almost a masterly skill in both plain and spherical Trigonometry, but to be well acquainted with the science of Navigation, and to have made great progress in Astronomy before he was removed to Oxford." Astronomy, mathematics, and navigation: a trinity which was to hold Halley entranced for the rest of his life.

This was an age when astronomy was being harnessed to the aid of navigation and the thorny question of finding longitude at sea. The heavens, it was hoped, would provide the accurate clock that appeared to be beyond the manufacturing abilities of clock makers. The hand of this nonmechanical clock would be the moon moving across the dial of the stars: the lunar-distance method of finding longitude. But this required an exhaustive measuring of the moon's erratic movement across the dial, a movement that Newton confessed gave him a headache when he thought about it. The measuring, calculating, and prediction of the moon's movement was a time-consuming operation (one to be measured in the eighteen years of the moon's saronic cycle) and one baited with traps, but the end product was a set of tables for the navigator to find his longitude.

The founding of the Royal Observatory at Greenwich (and thus the Greenwich Meridian) can be traced back to a French mistress of that most rakish of monarchs, Charles II. Louise de Keroualle, being Catholic, French, and avaricious, was heartily detested by her lover's subjects. For behind the beautiful baby face and the eyes that shed copious tears when thwarted, lay a calculating and mercenary mind.[1]

A compatriot of hers, she told Charles, had arrived in England with the solution to finding the longitude by means of the moon. He was introduced to Charles, who listened with interest. A Royal Commission to look into the Frenchman's claims was appointed. They, in turn, asked a young astronomer, the Reverend John Flamsteed, to advise them. His report was damning. Much more had to be learned about the moon's erratic movements to make the finding of longitude feasible. The Frenchman having lost his case, and to staunch his lover's tears, Charles signed a Royal Warrant to build an observatory: "Whereas, in order to the finding out of longitude of places for perfecting navigation and astronomy, we have resolved to build a small observatory within Our Park at Greenwich . . . with lodging room for Our Astronomical Observator and Assistant." A few days later, Sir Christopher Wren, the chosen architect; Robert Hooke, the inventor and physicist; Flamsteed; and the young undergraduate Halley were walking up the grassy hill to inspect the observatory's site. The foundation

---

[1]"Fubbs" was Charles's pet name for his mistress. But Nell Gwynn, having been ousted from the royal bed, always called her rival the "Weeping Willow."

stone was laid in 1675 and a year later Flamsteed started his work from the Royal Observatory's Great Room.

Flamsteed was to remain as Astronomer Royal, toiling away with his mural arcs, telescopes, sextants, and pendulum clocks (bought at his own expense, the royal munificence not running to instruments or his assistant's salary), gazing into the heavens and recording his findings, until his death in 1719.

At Oxford, Halley busied himself with geomagnetism, mathematics, and astronomy. He also published papers, remarkable for one so young, on the orbits of planets, sun spots, and the occultation of Mars by the moon. And then, with one year left to obtain his degree, he left Oxford to sail for the island of St. Helena in the South Atlantic. This remote island, run by the East India Company, he had selected as being the southernmost English possession where he could observe and catalog the southern stars to complement the northern star catalogs being produced by Flamsteed and other European astronomers.

It was an audacious undertaking, smoothed by the old-boy circuit and what today would be called networking. At the center of the net lay King Charles II. In the early September of 1676 the Secretary of State (and former provost of Halley's college) received a memorandum requesting:

Edmond Halley, student of Queen's College, Oxford, having been for some years a diligent observer of the planets and stars, has found it absolutely necessary, besides the continuation of observations here, that in some place between the Tropics, where the sun, moon and planets will pass near the zenith without refraction, their motions will be much better ascertained and navigation perfected, and that St. Helena will be a fit place, where the celestial globe may be finished, the stars in the southern hemisphere being very much out of their places. He humbly desires His Majesty's letter of recommendation to the East India Company that they will cause the ship ready to go to St. Helena to transport him and his friend thither, and that they may be received and entertained and have fitting assistance.

A month later the king instructed the East India Company to give Halley and a companion passage on their first ship bound for St. Helena. Within two weeks Halley and James Clark were boarding the *Unity* burdened by a dip compass; a magnetic compass; a pendulum clock; a huge brass sextant with a 5-foot radius and telescopic sights; a 2-foot radius quadrant; various telescopes, one 24-feet long; and a seaman's backstaff, designed by the Elizabethan navigator John Davis to measure the sun's meridian altitude, which Halley had improved with a lens. Halley was launched on his brilliant career.

St. Helena was something of a disappointment, being far too cloudy for all the observations required, and after a year Halley was back in England. Nevertheless, during the year on the island and the two voyages, he had observed the trade winds (to be used in his 1686 chart of the Atlantic wind system); the depression of the barometer at his observatory high on St. Helena's hills (to be used in another scientific paper); the shortening of his clock's pendulum (used by Newton as evidence that gravity was less in the low latitudes and that the Earth was not a true sphere but bulged along the equator). He had also made magnetic observations at sea for both dip—he found it zero, the dipping needle horizontal, at 15° north latitude—and variation.

His catalog of southern stars (the first catalog to be observed with telescopic sights and eyepiece micrometer) was published in late 1678. A copy of the planisphere, showing all these southern stars, was presented to Charles II and on it, with diplomatic astuteness, Halley had regrouped some stars into a new constellation which he had named *Robur Carolina*, a reference to the oak tree in which Charles had hidden after the Battle of Worcester during the Civil War.[2] The king, in return, asked Oxford University to give Halley his degree. Needless to say, it was granted. The same year, the twenty-two-year-old Halley was made a Fellow of the Royal Society.

His *Catalogue of the Southern Stars* and his energy in getting himself to St. Helena and building his observatory marked Halley, even to fellow astronomers, as a most unusual, brilliant, and practical man. Moreover, along with the academic brilliance, went a charm and self-deprecating sense of humor that won people's hearts and disarmed many a critic. He had, as the *Biographia Britannica* put it, "a vein of gaiety and good humour." All qualities he was going to need in full measure, when, seven years after diving for gold and ivory off the Sussex coast, he sailed as Captain Edmond Halley in command of a Royal Navy vessel to plot the variation of the compass and sail among Antarctica's icebergs.

[2]Oak Apple Day, or Royal Oak Day, on May 29, was a day when people wore sprigs of oak with gilded oak apples. It also happened to be Charles's birthday and the day he entered London at the Restoration.

Chapter 6

TO COMPASS THE GLOBE

In April of 1694 a small vessel just 64 feet long was launched on a high spring tide from the Royal Dockyard at Deptford on the River Thames. She was then entered into the Royal Navy as the pink *Paramore*.

Those were the days when vessels were designated by their hull shape, not their rig. The *Paramore*, with her high and narrow stern, shallow draft, and extreme tumble-home of topsides, was instantly recognizable, at least to a seaman's eye, as a pink. With apple-cheek bows and bulging sides, a pink looked as homely and as comfortable as a plump Dutch housewife. Which was very appropriate, for the name and shape came from Holland and pinks were used by the Royal Navy on the wifely duties of supplying stores and provisions. But the *Paramore* had not been built to succor the fleet, but to circumnavigate the world on a voyage of scientific exploration.

The *Paramore* and her voyage had been conceived a year before her launch. The proposal for the voyage had come from Halley and Benjamin Middleton, in a document of some three hundred words, asking the Royal Society to use its influence to obtain a small vessel "to compass the Globe from East to West through the great South Sea." The objective: to measure the magnetic variation, to fix the longitude of various ports and headlands, and to try the several methods of finding the longitude at sea proposed by some astronomers. Methods, the proposal hinted, that would be totally impractical at sea as they came from landlubber astronomers with no idea of the difficulties of taking celestial measurements from a moving deck.

It was an undertaking close to Halley's heart. A year before the proposal's submission he had published a paper on terrestrial magnetism with a detailed theoretical explanation for compass variation and the secular shift discovered by Henry Gellibrand. It was considered, in its day, one of Halley's most important publications.

After this proposal, events moved with remarkable speed— which suggests that hidden from the written historical record lay conversations and meetings over chines of beef, venison pie, oysters, gooseberry tart, ale, punch (if one were with Whigs), sack or claret (if one were with Tories), and brandy (if one were in mixed political company), all these meetings wreathed in tobacco smoke curling from long clay pipes.

The Royal Society approved the proposal; the Treasury smiled on "soe good and publick an undertaking" with its benefits for navigation and trade; Queen Mary II was enthusiastic; the Admiralty gave orders to the Navy Board; the Navy

Board gave orders to Fisher Harding, the master shipwright at Deptford. And so was born the *Paramore*.

The launching of the *Paramore* was followed by a mysterious two-year hiatus with the pink lying untried and untested. But with Louis XIV of France busily aggressive in Europe, North America, and the world's oceans, it was not, perhaps, the best of times to send out a small vessel on a scientific circumnavigation.

In the January of 1696 the *Paramore*'s torpor was disturbed by an inspection from the Navy Board. Five months later the Admiralty commissioned Halley as her Master and Commander, and issued warrants to a boatswain, gunner, and carpenter. But it was all a false start. Halley was appointed to the Mint at Chester to oversee the collecting of silver coins—which were being clipped by cheats and rogues—and to organize the issue of milled-edge coins. The *Paramore* went back into wet dock at Deptford.

HALLEY RETURNED TO London in 1698. Also arrived in London was Tsar Peter I of Russia and his entourage of drinking companions—the Great Embassy—who were touring through western European countries to study industrial techniques and enlist the services of men willing to work in, and modernize, Russia. In its efforts the Great Embassy had cut a boisterous, bibulous, and memorable swathe through Europe, which brought a weary wish from the Electress of Hanover that the manners of Peter and his party "were a little less rustic."

Peter, under the convenient incognito of Peter Mikhaylov, had come to England to spend his days studying shipbuilding at the Deptford yard, and his nights carousing at a tavern on Great Tower Street. The Russians lived at nearby Sayes Court, where their rustic manners blossomed. Sayes Court happened to be the diarist John Evelyn's home. He had sublet it to Admiral Benbow, who in turn had sublet it to Peter. In the three months that the Russians stayed at Sayes Court the lovely house and its gardens were reduced to ruins. Pictures were used as target practice; floors were ripped up and furniture destroyed for fires (it was a very cold winter, with the Thames freezing over); windows smashed; sheets and curtains torn; and Evelyn's carefully nurtured bowling green rutted and hedges broken down. Well might Evelyn's bailiff report to him that the house was full of people "right nasty."

The Tsar, a frightening bear of a man (he was well over 6 feet tall, with long powerful arms and, when under stress, an unnerving twitch which affected his face and caused his eyes to roll in their sockets until only the whites showed), had heard of Halley and the proposed *Paramore* voyage. Halley was invited to dine at Sayes Court and was quizzed by the Tsar, who found the Englishman good company with his conversation, wide-ranging knowledge, and fund of practical experience. Peter became so interested in the *Paramore* that he requested the Admiralty to rig and prepare her for a sailing trip. The Admiralty complied. And so the first time the *Paramore* went sailing was with a tsar at the helm. The Russians, much to the relief of their hosts, returned home in April. But a memory of this extraordinary visit still remains in Deptford: Czar Street.

THE PINK'S SAILING qualities proved poor. She was found to be crank (tender, in modern nautical parlance; easily heeled, in landlubber parlance) and the Navy Board was ordered to improve her sail-carrying powers. In August the Admiralty ordered the pink furnished with twelve months' worth of provisions and Halley was given his second commission as her Master and Commander. One of his first requests was for two azimuth compasses (for observation of the magnetic variation) and fishing gear. The Admiralty, meanwhile, had much correspondence with the Navy Board as to the number and nature of guns to be fitted aboard a craft outfitting for science—something that had never before crossed their desks, thus causing a bureaucratic problem. It was finally resolved at six 3-pounders and two pattereroes (very small guns on swivels).

On October 15, Halley was issued with his orders and instructions. No mention is made of a circumnavigation, the voyage being confined to the Atlantic for observation of the magnetic variation and the accurate positioning of ports and islands. Halley was also instructed to sail as far south as possible and search for the "Coast of the Terra Incognita, supposed to lye between Magelan's Streights and the Cape of Good Hope."

Seven weeks after receiving the Admiralty's instructions, Halley and the *Paramore*'s deck watch, a frigid northeasterly wind nipping at their ears, were eyeing the Devon coast as it vanished astern. Those seven weeks had been filled with delays and frustrations. The pink had proved a poor sailing vessel when beating to windward, and she leaked like a sieve. The sand ballast, with the constant pumping, kept choking the pumps. All this had forced Halley into Portsmouth, where the pink was recaulked and the ballast changed to shingle. The delays had set other worries to run through the minds of the crew. The lightly armed pink would be easy prey for the Moorish pirates—the Sallymen—operating from the North African coast. Fearful visions of being sold into slavery had led the crew to petition Halley to ask for an escort through the danger zone. A squadron, by pure chance, had been waiting off Portsmouth for a fair wind to sail down Channel and to the West Indies, via Madeira. The Admiralty ordered Admiral John Benbow to provide protection; and the Paramores had the satisfaction of watching four warships, totaling two hundred guns, as their escort into the Atlantic.

By December 21 the pink was sailing alone (having taken aboard wine at Madeira) and was heading for the Cape Verde Islands. The anchorage at Praia on São Tiago provided an unwelcome and hostile surprise: cannon fire from two English merchant ships. Luckily for the Paramores their compatriot's gunnery was poor; all their shots fell short or wide. The *Paramore* was hove-to and Halley sent a boat to ask why they were firing at the "King's colours." They had mistaken her for a pirate, came the reply.

This bizarre episode was to set the tone for Halley's first voyage in the *Paramore*. Six months after being fired upon, followed by a clockwise sweep across the Atlantic to Brazil and the West Indies, the pink was back in England, anchored in the Downs among a naval squadron commanded by Admiral Sir Cloudesley Shovell.

After riding from Deal to London by post-chaise, Halley explained to the Admiralty the cause of his early return. His voyage, he told Their Lordships, had been cut short for two reasons, the first being the lateness of the season for any exploration deep into the South Atlantic; the second being the almost mutinous behavior of his only other commissioned officer, Lieutenant Edward Harrison. This officer's animosity and stirring-up of trouble with the crew had made the voyage so "very displeasing and uneasy" that at one point Halley had been forced to confine Harrison to his cabin. But on the plus side a start had been made on the measurements for magnetic variation.

The Admiralty could not ignore Halley's accusations against Harrison and some of the warrant officers for their "abusive language and disrespect." A court-martial was ordered, presided over by Shovell. The court, much to Halley's chagrin, found no officer had disobeyed Halley's orders, although "there may have been some grumbling among them as there is generally in small vessels under such circumstances." The officers, however, were given a severe reprimand. And, during the proceedings, Harrison's malice toward Halley came to light.

Four years previous to being commissioned to the *Paramore*, Harrison had sent a paper to the Royal Society on methods of finding the longitude at sea. Halley had read the paper to the Council where it had been dismissed as saying nothing new on the subject. This dismissal had festered in Harrison's mind. Two years later, in 1696, he published his expanded paper as a book, *Idea Longitudinis*. Copies were sent to the Admiralty, the Navy Board, and the Royal Society, where it met with the same reception as his paper. The book reeks of an unbalanced personality. He is vituperative about mathematicians and those "Persons in *England* whose Duty it is (being paid for it) to improve *Navigation* and *Astronomy*, and from which much is expected, and little or nothing appears." As he had served seven years at sea in merchant ships and six ships of the Royal Navy, he thought himself "to be a more Competent Artist in

Navigation" than any mathematician. Admitting that his knowledge of mathematics was small, this navigational ability rested entirely on "Divine Authority."

Ironically, considering the court-martial charge brought against him, the Dedication of his book reads: "It is a saying in the *Navy*, He that knows not how to obey *Command*, is not worthy to bear command. . . . As it is the Duty of a Subject to be True and Loyal to his Prince, so it is the Duty of Servants, to be Faithful, Humble and Submissive to their Masters."

But irony is laid on irony. For it was Halley who had requested a commissioned officer for the *Paramore* (he had small faith in the qualities of his warrant officers) to ensure discipline. The Admiralty had given him Harrison, a man who had nursed his grudge over the years and who blamed Halley for the poor reception given to both paper and book. And who now found himself under the command of the loathed (and unsuspecting) mathematician and astronomer.

Halley, the voyage cut short, thought the court's verdict too lenient, and voiced his opinion in a letter to Josiah Burchett, the Admiralty secretary. But legally, as the charge had been insolence and not disobedience, the court could do nothing but hand out reprimands.

Harrison resigned his commission and vanished into the merchant marine. Halley, back in London, attended the regular meetings of the Royal Society where he showed plants collected on the voyage and a chart plotted with his observations of magnetic variation. London's gossipers, their tongues wagging in the coffee houses, had it that Halley had been lucky to return alive, as the *Paramore*'s crew had obviously intended to turn pirate.

The Admiralty, however, had not lost faith in Halley and

his determination to continue his magnetic observations. Orders were issued and the pink was prepared for a second voyage into the Atlantic with Halley as her commander. This time he was the only commissioned officer. But the Navy Board, in their infinite wisdom, gave him a one-armed boatswain to replace the mutinous boatswain, which caused Halley to point out that a one-armed boatswain would "be a little service in case of extremity." And, knowing the Admiralty and their obdurate ways, he requested three or four extra hands to make up for the single-handed boatswain.

Halley's second attempt at the Atlantic to chart magnetic variation was off and running.

Chapter 7

HALLEYAN LINES

In the autumn of 1699 the *Paramore*, ten months after sailing on her first voyage escorted by Admiral Benbow's squadron, was once again sailing down Channel, but this time accompanied by the *Falconbird*, a Royal African Company ship bristling with thirty guns, as her insurance against the predatory Sallymen.

A few days later, having left the chops of the Channel and now some 200 miles off the Spanish coast, Halley made an entry into his journal which highlights a hazard endemic to that age of ocean sailing. Just as land travelers were menaced by brigands, bandits, footpads, and highwaymen, the sea traveler always cocked a questioning and suspicious eye at every topsail that lifted above the horizon: enemy man-of-war, privateer, picaroon, pirate? A fleet of Danish merchant ships, homeward bound, warned the *Paramore* and the *Falconbird* of a

Sallyman lying across their course. The sail turned out to be a lone English sloop. Next morning they saw another sail and hove-to to examine it through spy glasses. It proved to be a Dutch vessel (which hoisted French colors) bound from Faro with a load of figs. She was, as an envious Halley put it, comparing her to his homely *Paramore*, "exceeding sharp and promised to Saile like the Wind."

The *Paramore* and the *Falconbird* parted company a week after those suspicious, but innocent topsails had lifted above the horizon, the *Falconbird* bound for the African coast and the *Paramore* for Madeira to load wine. But contrary winds blocked the *Paramore*'s approach, and Halley dryly noted that that his crew chose to go without their wine rather than spend time struggling to windward in an area where they were in danger of being taken by the dreaded Moorish Sallymen.

The same day, in heavy seas, Halley's servant ("my poor boy Manley White") fell overboard and drowned. It was a tragic accident that deeply affected Halley. The telling of this calamity, and of the vain rescue attempts in the rough seas, always brought tears to Halley's eyes.

After watering and buying provisions in the Cape Verde Islands the *Paramore* sailed south across the Equator. Before sailing, Halley wrote a letter to the Admiralty (to be carried by a homeward-bound English ship) telling of his progress and that the "ships company is all well and my officers as forward this time to serve me, as they were backward the last." He ended his letter by saying that he hoped to reach the southern limits of his voyage by the New Year.

But on January 1, 1700, the *Paramore* was only a few miles south of Rio de Janeiro. It had proved a slow and tedious passage across the South Atlantic with a foul bottom and light winds. Two weeks had been spent in Rio de Janeiro preparing for the voyage south and loading casks of rum: the lack of Madeira wine was obviously making itself felt.

A month later, 1,600 miles south of Rio, Halley found the temperature in his cabin close to freezing. A few days later they were surrounded by porpoising penguins and the temperature in Halley's cabin fell below freezing. Then the Paramores came upon the voyage's most amazing sight: three tabular icebergs which Halley estimated at two hundred feet high, with one five miles long. Two of them were named, the white cliffs of ice reminding them of familiar English headlands, Beachy Head and North Foreland. Halley sketched them into his journal: the first drawings ever made of Antarctica's tabular icebergs—the massive flat-topped icebergs spawned from the ice shelves surrounding the continent.

The next days brought strange and frightening hazards to the Paramores—thick fog and strong winds, with the pink sailing a blind and terrifying slalom course between icebergs and pack ice. On February 7, Halley decided to quit this awful area "to recover the warm Sunn, whom we have Scarce seen this fortnight." For ten days, with no sun and poor visibility, any compass variation measurements had been impossible. But progress north was slow, for now came head winds with dark moonless nights. And the fear of icebergs, lying like footpads in their path, forced them to heave-to at night. By mid-February, having a desire to see the remote islands of Tristan da Cunha, which he estimated lay some 360 miles away, Halley set a course to find these isolated clumps of volcanic rock set halfway between South America and South Africa. Less than a week later, right over the *Paramore*'s bow, the island peaks hoisted themselves above the horizon. It was a pretty piece of navigation by Halley. This was not the usual method of finding your destination, which was to sail into its latitude and then sail along it—known as running the latitude down. Halley had found them by sailing at a diagonal across the lines of latitude to come upon his objective.

The Tristan da Cunha group lie close to the latitude of the Cape of Good Hope, and it was toward the Cape that Halley now set his course to wood and water. Halfway to the Cape they were struck by hurricane-force winds, the small pink being driven north before the storm with as much chance of steering a course as a leaf torn from a tree. The *Paramore* broached in the mountainous seas, and with decks awash, the water swirling knee-deep into Halley's cabin, she came close to foundering. But, in the heartfelt words of Halley, "it pleased God she wrighted again."

Driven so far north of their course to the Cape, with their provisions ruined by salt water, Halley now decided to sail for St. Helena. But this time he was going to find the island by the conventional method of running its latitude down. On March 11, the island appeared over the horizon. Hours later they were anchored off Jamestown and Halley was soon ashore at the familiar island, meeting old friends and organizing water and fresh provisions for his men after ten uncomfortable, and sometimes terrifying, weeks at sea.

The *Paramore* sailed at the end of the month, bound across the Atlantic for Trindade Island off the Brazilian coast. Before sailing, Halley left a letter, to be carried by the next homeward-bound East Indiaman, for the Admiralty. The last sentence in the letter claimed that he had found a viable theory for compass variation.

At Trindade, they watered (getting rid of most of the St.

Helena water, as it was barely drinkable) and Halley put ashore some goats, pigs, and guinea hens for the succor of any shipwrecked sailors. Leaving the Union Flag flying over the native land crabs and colonizing goats, hogs, and guinea fowl, the *Paramore* sailed for Brazil.

At Recife the amiable Portuguese governor gave permission to provision and buy wine. Halley was also relieved to hear that Europe was at peace. But Halley's peace of mind was ruffled by the behavior of a certain Mr. Hardwick, who claimed to be the English consul. Refusing to believe Halley's commission, he kept the furious astronomer under guard at his house, and then promptly searched the *Paramore* under suspicion of being a pirate ship. Halley later learned from the Portuguese that Hardwick was no consul but an agent of the Royal African Company looking to line his own pocket by seizing the *Paramore* as a prize.

Seventeen days after sailing from Recife, with Halley still fuming on Hardwick's conduct, the *Paramore* was anchored off Barbados in the West Indies. They found the island in the grip of a deadly sickness. The governor's advice was succinct: Leave as soon as possible. But the two days loading fresh water were enough for Halley and some of the crew to become seriously ill. Sailing to St. Kitts and then Anguilla to load more water and wood brought recovery and relief. But Halley thought the sick had only recovered due to the "extraordinary care of my Doctor."

The *Paramore* now sailed north to Bermuda, arriving after an uneventful passage with a healthy crew but much weathered ship. The tropic sun had shrunk her deck planks and the tropic seas had produced a fine crop of weeds and barnacles. For the final stage of the voyage home the pink's deck was recaulked, the parched wood given a coat of paint, the hull careened, and the bottom scrubbed clean.

The spruced-up vessel sailed from Bermuda and, aided by the Gulf Stream, steered for Newfoundland. Three weeks later they were groping their way through thick fog, the leadsman calling out the depths, as they approached the coast. The *Paramore*, creeping through the fog, startled some English vessels fishing for cod, who fled away like chickens before a fox. At anchor in Toads Cove a Devon fishing boat fired several shots through their rigging. But the whole matter was cleared up when the skipper explained that a pirate ship had been marauding and plundering along the Newfoundland coast but a few days before.

The Atlantic crossing from Toads Cove took three weeks. On September 9, 1700, Halley handed the *Paramore* over to the Royal Dockyard at Deptford.

HALLEY'S TWO VOYAGES, considered by many to be the first voyages for a purely scientific purpose, had finally ended. Halley, ever a convivial man, was soon meeting with old friends at the King's Arms on Ludgate Hill and working on the presentation of his findings to the Royal Society. From the mass of magnetic variations, and in remarkably short time, he produced a manuscript chart and showed it to members of the Society. It was a chart "curiously laid down with Marks," according to the Royal Society Journal Book.

The marks—which Halley called "Curve Lines" and his contemporaries "Halleyan Lines"—became famous. Halley had joined the points of equal magnetic variation with a line. It was a brilliant cartographic concept and one that has

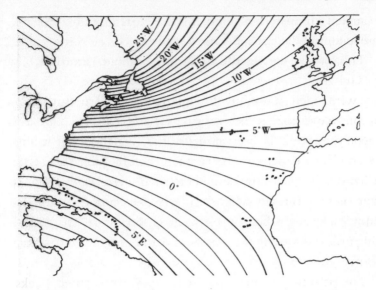

Variation chart of the North Atlantic for the year 1701. Taken from Edmond Halley's Variation Chart.

become standard for geographic observations. The modern map depicts the same idea for equal heights with contour lines; a weather map for equal barometric pressures with isobars; equal temperatures with isotherms; equal depths below sea level with isobaths. On today's magnetic variation charts the lines are known as isogones.

In June of 1701 a printed version of this map was produced. On it Halley explained his curve lines:

The **Curve Lines** which are drawn over the Seas in this Chart do shew at one View all the places where the **Variation** of the **Compass** is the same. The numbers to them shew how many degrees the **Needle** declines either **East-**

wards or **Westwards** from the true **North**; and the Double Line passing near **Bermudas** and the **Cape Verde** isles is that where the **Needle** stands true, without **Variation**.

In his "Description" to the chart Halley warned mariners of the secular changes to compass variation: "This Chart, as is said, was made by Observations of the year 1700, but it must be noted, that there is a perpetual, tho slow Change in the variation almost everywhere, which will make it necessary in time to alter the whole System." He then gave his estimated variation changes for the Cape of Good Hope, the English Channel, and the Atlantic coasts of Africa and South America.

The following year Halley published a World Chart of

Variation Chart of the North Atlantic for the year 2001, showing the change in variation over the three hundred years since Halley's chart.

compass variation. It was a chart destined to become a mariner's favorite—Captain James Cook used a copy—and one that went through many editions and revisions, some foreign, and was still being used at the end of the eighteenth century.

HALLEY'S CONNECTION WITH the *Paramore* continued. In 1701, he was once again in command, this time on a survey of the English Channel coast, with particular attention being paid to tides and compass variation. The resulting chart shows depth in fathoms, and current direction by arrows. Sprinkled across the chart are Roman numerals which, when applied to a formula using the phases of the moon, gives the time for high water. Compass variations are also shown. Halley, with his tidal chart, was once again leading the field. Over a century would pass before any other tidal chart, for any other sea, was published.

Halley also published a pamphlet on the grave dangers of approaching or leaving the English Channel if the latest changes in compass variation were neglected. In *An Advertisement Necessary to be Observed in the Navigation Up and Down the Channel of England*, he warned ship masters of two hazards which could lead to shipwreck and death. On most charts the Scilly Islands were laid down too far north of their true position, which was hazard enough. He also warned that compass variation had changed in a few years from easterly variation to 7½ degrees westerly variation. Masters neglected this change at their peril. "If they miss an Observation for two or three Days, and do not allow for this Variation, they fail not to fall to the Northward of their Expectations." In short, a master approaching the English Channel, and running his latitude down (Halley recommended nothing north of 49°40') would, for every 80 miles sailed, be 10 miles north of his dead-reckoning position.

Six years after publishing this pamphlet, Halley, now a professor at Oxford, heard the shocking news of Shovell's fleet being wrecked upon the Scilly Islands. Some 250 years later, Commander W. E. May, R.N., inspected the surviving log books and found that no allowances had been made for compass variation.

Ships sailing down Channel could also be led into a trap. Hidebound masters, still steering what was known as the "Channel Course"—WSW—from Beachy Head would, not allowing for the westerly variation, inevitably pile up on the coast of France or the Casquets off the Channel Islands. Indeed, the growing number of ships being wrecked upon the Casquets and the French coast could be cited as proof of this. Halley suggested, allowing for the westerly variation, a safe compass course of West-by-South when taking a departure from Beachy Head.

Halley's naval career came to an end with his English Channel survey. But a robust residue of his days at sea must have remained. A censorious Flamsteed, hoping to block Halley's appointment as an Oxford professor, sourly noted that Halley "now talks, swears, and drinks brandy like a sea-captain." Flamsteed's efforts failed.

Flamsteed died sixteen years later. Racked by gout and migraine headaches, England's first Astronomer Royal had spent those years in bitter arguments with the Royal Society, Sir Issac Newton, and Halley. The second Astronomer Royal, ironically enough, was Halley. His first job was to equip the

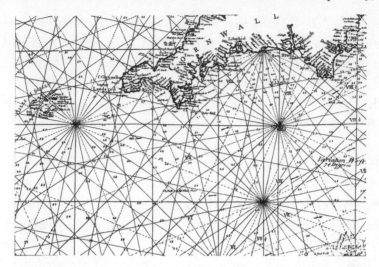

Portion of Halley's chart of the English Channel showing the loxodromic lines and the compass variation marked as 7½ degrees.

Royal Observatory with instruments, for Flamsteed's wife had stripped the Greenwich Observatory clean.

Halley's second project was to set in motion an undertaking dear to his heart, the study of the moon's cycle in the heavens, a cycle, known as the saronic cycle, which lasted eighteen years. For a man aged sixty-five it showed a touching faith in his own longevity. But the results, once tabulated, would provide a method for finding the longitude at sea.

Halley's faith in his longevity was realized. He died in 1742, aged eighty-five, peacefully and quietly after drinking a glass of wine. He was buried next to his wife in a churchyard close to the Greenwich Observatory.

One of Halley's earliest contributions to the Royal Society's *Philosophical Transactions* had been on compass variation. It seems appropriate that fifty years later, in 1732, one of his last contributions concerned the same subject.

In 1756, Mount and Page of London, publishers of Halley's world variation chart, brought out a new and improved edition showing the changes from the original. A pamphlet stated that the greater number of measurements and their accuracy were due to new improvements in the magnetic compass. For this essential instrument had been neglected until "the judicious Dr. Gowin Knight, F.R.S., examined into their fabric and construction and employ'd his Magnetic Knowledge towards their Improvement."

But Dr. Knight's improvements were to be disputed by some very famous and practical seamen.

Chapter 8

DR. GOWIN KNIGHT
AND HIS
MAGNETIC MACHINE

North Atlantic winter storms can be fearsome affairs. On one such night, loud with thunderclaps, howl of wind in rigging, roar of breaking seas, and with St. Elmo's fire flaming eerily at the mastheads, the ship *Dover* was struck by lightning. So violent was the strike that most of her crew, including the captain, were temporarily blinded. The lightning bolt drove through two decks and opened up planks on the starboard side. Within fifteen minutes, nine feet of water was sloshing around in the bilges. The storm lasted for five days and the catalog of disasters continued with the loss of two masts and most of the sails.

Included among these horrors was the lightning's effect upon the compasses. All were demagnetized so as to be utterly unserviceable. The *Dover* had been some 600 miles to the west of the Scilly Islands, on a passage from New York to London, when struck by the thunderbolt. Twelve days later, on January 21, 1749, with pumps working overtime and having steered their course by the wind, sun, and stars, the shattered, jury-rigged *Dover* and her tired crew anchored off Cowes in the Isle of Wight.

The *Dover*'s dry compasses were typical of their day: a decorative card with the compass needle hidden underneath, the card balanced and pivoting on a vertical pin. Three compasses had wooden bowls and one a brass bowl; all the bowls were fitted with a glass top. The compasses were mounted in wooden boxes and swung in gimbals, with the main steering compass fitted into a binnacle. Binnacles of those days were large wooden boxes fitted forward of the wheel. They usually contained three compartments divided by glass. The middle compartment was for the compass, the side compartments for an oil lamp and a sand glass for timing the speed. On some larger ships the end compartments held compasses: the helmsman having to stand beside the large steering wheel, the middle compartment being reserved for the oil lamp or candle. A vertical line, the lubber's line, was marked on the inside of the compass bowl. This line, most probably, was a sixteenth-century Portuguese innovation, the line representing the fore-and-aft centerline of the ship, thus enabling the course to be steered by bringing the lubber's line to the desired point on the compass card.

At Cowes the *Dover*'s skipper, Captain John Waddell, bought a compass to get his ship to London. And found yet another problem. Placed inside the binnacle the compass card swung two points (22½ degrees) from the direction it stood when outside the binnacle. He then moved the empty binnacle to a different part of the deck. Every time the compass was placed inside the binnacle the card repeated its swing. Iron, thought Waddell, could not be a possibility as he had given strict instructions to the shipwright that no iron nails be used in its construction.

Having safely navigated to London, Waddell decided to investigate the puzzling compass behavior by asking advice from experts. And the London of 1749 had just such an expert: Dr. Gowin Knight.

THE THIRTY-SIX-year-old Knight, like the Elizabethan Dr. William Gilbert, was a physician. But unlike Gilbert, the man of ample means who could afford to indulge his magnetic passions, Knight was a man of slender means who sought to become wealthy *by* his magnetic passions.

The fascination for the lodestone's hidden power and the fashion for terrellas had not waned since Gilbert's day. Two of the Royal Society's proudest possessions were the terrellas once owned by Sir Christopher Wren and the Earl of Abercorn. In 1730, Abercorn had drawn up a table for calculating the monetary value of a lodestone based on its size and magnetic power. Sir Issac Newton had a tiny chip of lodestone, weighing 3 grains, set in a finger ring which could support a weight of 700 grains. Lodestones were used in entertaining party tricks; and no collector's cabinet was complete without a lodestone, shaped into a rectangular block and mounted in a silver case. Instrument makers sold lodestones. One London instrument maker's trade card, now in London's Science Museum, illustrates his wares: globes, quadrants, telescopes, sundials, parallel rules, spectacles, thermometers, compasses, and a lodestone demonstrating its power by holding up an anchor. Thomas Tuttell, another London instrument maker, also sold mathematical playing cards with the seven of spades delineated as the "Magnet or Loadstone" card. Shown on the card are miners digging for lodestone, a lodestone holding up a heavy key, another lodestone holding up Mahomet's tomb (according to ancient legend, the tomb was made from iron and held suspended in midair by a powerful lodestone), and an anchor (the standard image for all things nautical), illustrating the lodestone's vital use in navigation for its power to remagnetize compass needles. The card also contains a potent message on the magnet's power: "A treasure of hidden virtues which has made our **Navigation** great, our **Comerce** general, our charts & **Globes** much more accurat & exact."

It was in this commercial London of the 1740s, with its emphasis on lodestones and magnets aiding navigation and trade, that Knight started his practice as a physician. Born in 1713, the son of a provincial clergyman, Knight had been educated at the Leeds Free Grammar School and Oxford's Magdalen College.[1] In the college's somnolent academic atmosphere, with its library books covered in mold, Knight studied medicine and dabbled with the study of magnetism and the making of artificial magnets. How or why Knight started his magnetic experiments is something of a mystery. But a year before he arrived at Oxford, Servington Savery's paper on magnetism had been published in the Royal Society's *Philosophical Transactions*. Savery's son happened to be at Oxford with Knight.

Once settled in London the thin-lipped, beak-nosed

---

[1]The same college castigated by Edward Gibbon: "I spent fourteen months at Magdalen College; they proved the fourteen months the most idle and unprofitable of my whole life."

Knight, a man very conscious of his own dignity and rightful place in the world, promptly set about attracting the great and the good to witness the powers of his artificial magnets. As the Royal Society represented the weighty and prestigious arbiter of eighteenth-century natural philosophy, Knight made the very astute move of inviting its president, Martin Folkes, to his lodgings. On two visits in 1744, with the dank, sooty November mists pressing against the windowpanes, Knight demonstrated with adroit showmanship the amazing powers of his artificial magnets, so much more powerful than any lodestone. Folkes was dutifully impressed and asked Knight to repeat his performance at the Thursday evening gathering of the Royal Society in its rooms at Crane Court, just off Fleet Street, and close to Knight's lodgings. Here he succeeded in convincing the assembled Fellows that he had invented a new and powerful method of making magnetized steel bars.

Knight now set about collecting men of influence about him with the same efficiency as one of his new magnets collected iron. He read papers before the Fellows on magnets and magnetism; was elected a Fellow, and then, in 1747, was presented with the Copley Medal, the Royal Society's highest award. At the award ceremony Folkes told the bewigged Fellows of how Knight had "applied himself very particularly to consider the best methods of touching needles for the sea service," and thus had "contrived an artificial apparatus and that very portable and convenient, by which as strong, if not stronger touch may readily be given to needles than any they could receive from the best natural loadstones." Warming to his work and in full grandiloquent flow (gathering in Euripedes, Plato, and Aristotle on the way), he ended by saying that the lodestone—and by inference, Knight's superior magnet—by magnetizing compass needles had "proved of the most extensive use and benefit to mankind: and it is by the help of that chiefly that we have been enabled to make, with security and ease, long voyages by sea; and consequently to increase and promote greatly our foreign trade and commerce whereby we are provided at home with the fruit, the convenience, the curiosities and the riches of the most distant climes."

Knight, with the Copley Medal, had scaled the natural philosopher's greatest heights. A year later he was elected a member of the Society's exclusive Thursday evening dining club, a membership considered by the clubbable as more important than the Copley Medal. The club, restricted to forty members, met at the Mitre Tavern on Fleet Street at four o'clock in the afternoon, four hours before the Society's evening gathering. The tavern, very conveniently, was but a few minutes walk from the Society's rooms. Just as well, perhaps. A typical meal provided by the tavern's keeper ran along the following lines: fresh salmon, fried cod, cod's head, boiled chicken, haunch of venison, bacon, pigeon pie, chine of house lamb and watercress, spare ribs of roast pork with apple sauce, plum pudding, apple pie, butter and cheese—all washed down with claret, port, and brandy.

The doctor was now the Society's expert on all matters magnetic and, most satisfyingly, was making money by selling his magnets. But he steadfastly refused to divulge his secret manufacturing method: this would have been killing the goose laying those golden magnetic eggs. Such secrecy brought criticism from some Fellows and Dr. Samuel Johnson, all of them united in a belief that a natural philosopher's findings should benefit mankind rather than the philosopher's wallet. Knight refused to budge on his policy of secrecy. But

Knight had rivals in the field of making steel magnets. And one of them, John Canton, demonstrated his method before the Fellows and then published it. Knight was furious.

It was not until after his death in 1772 that his friend and executor, Dr. John Fothergill,[2] the Quaker physician, botanist, and philanthropist, revealed the Knight method. The original starting point had been time-consuming. With a natural lodestone, Knight magnetized, by stroking, a number of steel bars. These, when secured together, formed a magnet much stronger than the original lodestone. With this magnet he repeated the process on more steel bars to make an even stronger magnet.

He then made life easy for himself by constructing a magnetic machine to mass-produce magnets and compass needles. This contraption consisted of two wheeled magazines, each weighing over 500 pounds. Each magazine was made up of 240 bar magnets tightly bound together. The two magazines could be wheeled close to each other and a steel bar or compass needle placed between them, the result being a strongly magnetized bar or compass needle.

To sell his product, Knight had to tread a fine line of salesmanship between his gentlemanly status as a physician, complete with grave air, silver-topped cane, and full-bottomed wig, and the more colorful salesmanship of the many flimflam entre-

---

[2]Knight was very much in Fothergill's debt. An unwise investment in a Cornish mining venture, he told Fothergill, had put him into financial difficulties. Visions of bum-bailiffs, sponging-houses, and debtor's prisons were eighteenth-century nightmares. A sum of one thousand pounds, said Knight, would banish these nightmares and make him happy. "I will then make thee happy" replied Fothergill. And promptly gave him the money.

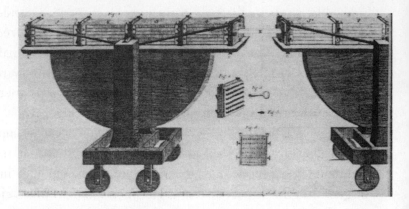

Dr. Gowin Knight's magnetic machine for making artificial magnets. Each one weighed over 500 pounds. From *Philosophical Transactions,* vol. 66 (1776).

preneurs of the eighteenth century, for this was the glorious age of mountebanks, merry-andrews, charlatans, and quack doctors. The age of potions, pills, powders, and nostrums whose bizarre ingredients would find a happy home in the bubbling contents of a witch's cauldron: worms, powdered snails, soap, urine, eggshells, stewed owls, fish eyes, antimony, arsenic, carrot seeds.

The arch-quack of the eighteenth century has to be James Graham, a Scot who came south but died in Edinburgh, a lunatic. Graham's enterprise relied not so much on potions and pills but on the beneficial effects of magnetism and electricity. From his "Temple of Health" on the Royal Terrace facing the Thames, and then from a house in Pall Mall, Graham catered to the tastes of the rich, jaded, sterile, and impotent. The centerpiece of his extraordinary enterprise was the "Celestial Bed" for curing the impotent and those wishing to conceive the perfect child. This facility cost £50 a night. Here, Graham's clients grappled in a perfumed bed 12 feet long by 9 feet wide, domed and lined with mirrors, on a silk-covered mattress stuffed with "strongest, most springy hair, produced at vast expense from the tails of English stallions." A small ensemble played soft music to the couple. And, as if this were not enough to arouse the jaded, nearly a ton of magnets underneath the bed poured forth their "exhilarating force."

Knight was subtler and less flamboyant in his salesmanship. As a Fellow of the Royal Society the initials F.R.S. after his name added a certain gravitas, and subtly suggested that his magnets had the Society's blessing. His publications in the Society's widely read *Philosophical Transactions* formed a free, and prestigious, way of advertising. He added to his credentials by publishing a deeply impenetrable book on the cosmos.[3] A third of this tome was given over to magnetism. A rueful Benjamin Franklin, who sat at many a Royal Council meeting with Knight, although admitting that Knight was the "greatest master of practical magnetics that has appeared in any age," having grappled with Knight's glutinous prose, cheerfully admitted that he had never found the "leisure to peruse his writings with the attention necessary to become master of his doctrine."

But Knight the entrepreneur, having inspected the *Dover's* compass and binnacle, saw an opportunity to expand from the selling of his magnets for the remagnetizing of compass needles. He knew his steel magnets were the best. They should be used in conjunction with the best of compasses. And he would design and build this best of compasses.

[3] The title gives more than a hint of its flavor: *An Attempt to Demonstrate That All the Phenomena in Nature May Be Explained by Two Simple Active Principles, Attraction and Repulsion: Wherein the Attraction of Cohesion, Gravity, and Magnetism, Are Shewn to Be One and the Same; and the Phenomena of the Latter are More Particularly Explained.*

Chapter 9

KNIGHT'S COMPASS

Knight's inspection of the *Dover*'s compass and binnacle revealed a sorry tale of shoddy workmanship and dishonesty. The sides of the compass box, Knight found, were pinned together with iron nails, sixteen on the sides and ten on the bottom. All had been magnetized by the lightning strike. The binnacle, which Waddell had ordered to be free of iron nails, had four large iron spikes in its construction, enough iron, thought Knight, to affect even the best of compasses. And the *Dover*'s compasses, even with the most elastic of imaginations, could not be considered the best.

As was normal for the times, the needle was made up from two magnetized wires separated at the middle and joined at the ends, forming a diamond shape. This was glued to the underside of the compass card and a brass cap for the vertical pivot pin fitted at the center of the diamond shape formed by the wires—except that this diamond was misshapen and the brass cap offset from the center. The lightning had also reversed the polarity of the diamond-shaped needle. Such a compass, fulminated the appalled Knight, could only be considered a "despicable" instrument. "How then," continued Knight, emphasizing his anger with a liberal scattering of exclamation marks, "must any one be shocked to hear that almost all the compasses, made use of by our trading vessels, are of the same sort! The boxes all joined with iron wire, and the same degree of inaccuracy observed throughout the whole."

This state of affairs convinced Knight that he should, could, and would, design a compass on proper magnetic principles.

THE LONDON THAT Knight lived in, with its Hogarthian images of drunks, rakes, and prostitutes, seems an unlikely city for the making of precision instruments. But between the squalid alleys and courts to the east of the Tower, where lived seamen, porters, pawnbrokers, footpads, and highwaymen, and the rapidly growing West End, with its open squares, paved streets, broad pavements, and piped water, lay a huge metropolis, the largest city in Europe. A glance at the names given to London's streets, alleys, and squares gives a pungent whiff of its diversity. In the East End: Black Dog Alley, Harebrain Court, Shoulder of Mutton Alley, Little Pump Yard, Brew House. In the West End, the pungency gave way to the cambric handkerchief and powdered wigs: St. James's Square (in 1721 home to six dukes, seven earls, one countess, one baron, and one scrub of a lowly baronet. George III was born there), Jermyn

Street, Grosvenor Square, Berkeley Square, Cavendish Square, Hanover Square, King Street, Duke Street, Duke of York Street, Albemarle Street, Charles II Street.

In between these two poles lay a patchwork quilt of trades employing thousands of apprentices, craftsmen, journeymen, artisans, and laborers. The river and port alone employed a quarter of London's workforce. Thousands were employed in silk weaving in Spitalfields, sail-cloth weaving in Tower Hamlets, porcelain factories in Bow and Chelsea, coach and furniture making in the Covent Garden area, watch and clock making in Clerkenwell and the neighbouring parish of St. Luke's. These two parishes also produced cutlery and surgical, optical, mathematical, navigational, and surveying instruments.

From this pool of skilled craftsmen came the men who made the instruments that had made London the envy of scientific Europe. Men such as John Bird, who had made the 8-foot-radius quadrant for the Greenwich Royal Observatory, a machine of such accuracy that he was soon making copies for observatories in France, Germany, Russia, and Spain. And George Graham, the Cumberland-born Quaker who supplied the French Academy with the apparatus used for the measurement of a degree of the meridian, and whose invention of the "dead beat" escapement and mercurial pendulum in clocks was to be used in observatories throughout the world for over 150 years. Graham, unlike Knight, was very frank in communicating his discoveries. A generous man with a conscientious objection to interest, he kept his money in a strong box and loaned large, interest-free amounts to friends. One of the men to benefit by this generosity was a fellow clock maker, John Harrison. Graham, as a Fellow of the Royal Society, contributed papers to the *Philosophical Transactions* on compass dip and variation. He also made a delicate instrument to measure the diurnal changes in terrestrial magnetism. Thomas Mudge, one of Graham's apprentices, invented the lever escapement, made Dr. Samuel Johnson's first watch, and then devoted himself to making nautical chronometers. Jesse Ramsden was the inventor of a remarkable "dividing engine" for graduating precision instruments. This was a machine that reduced the size of sextants by a half with no loss of accuracy. Ramsden, the son of a Yorkshire innkeeper, produced thousands of mathematical and nautical instruments and employed sixty craftsmen. He was a man of simple tastes, happiest when sitting by his kitchen fire and drawing plans, a cat on one side, a mug of porter and a plate of bread on the other. Apprentices would sit around as he drew, whistled, and sang. After he explained a design he would say: "Now, see man, let us try to find fault with it." And if a completed instrument fell short of his ideal it was destroyed, with the exclamation: "Bobs, man! This won't do. We must have at it again."

KNIGHT LACKED THE companionable ease of Jesse Ramsden. But, like his fellow Yorkshireman, he had tenacity in teasing out problems.

The *Dover's* shoddy compasses led Knight to ask an important question. Were all marine compasses so ill-made? Twenty compasses were collected and Knight set about his analysis. Most of the needles, he found, were made from iron wire bent into a lozenge shape. The wires had only been hardened at their ends by being heated to red hot and then quenched in water. This gave the lozenge shape a measurable magnetic imbalance—to such at extent that none of the compasses agreed as to the direction of magnetic north!

Royal Navy vessels and some merchant ships had compass needles made from one length of tempered steel. Broad at the ends, they tapered toward the middle where a hole had been drilled for the pivot-pin cap. The ends of these needles were shaped according "to the skill or fancy of the workman." But these needles, thought Knight, had to be considered preferable to those lozenge-shaped wire abominations. They still, however, had their oddities. Magnetic sand, sprinkled across a sheet of thin glass set over the needles, showed them to have six poles instead of the expected two: one pole at each end, two where the needle tapered, and two at the pivot hole in the middle. But two of the needles, which were straight and square at the ends, were found to have only two poles and only small disturbance of the magnetic sand at their pivot holes.

From this Knight concluded that a long rectangular needle with no pivot hole was the best and simplest. This needle would be placed *above* the compass card. The card would be made from very thin, varnished paper supported at its circumference by a brass ring. The ring would thus counterbalance the needle's weight. The pivot cap would be fitted to the underside of the card, thus eliminating any pivot hole in the needle. To reduce friction in the pivot cap, after various experiments with different materials, he decided upon a cap turned from ivory fitted with a piece of agate. The pivot pin, in contrast to this elegant and expensive solution, was a common sewing needle fitted in such a way as to be easily replaced. This rather plebeian solution was chosen because "they are generally better pointed than any that a common workman could pretend to make extempore."

IT WAS NO common workman who worked with Knight and made the doctor's first compass. John Smeaton, aged twenty-six and eleven years younger than Knight, had been born in Leeds and, like Knight, had attended the Leeds Free Grammar School. Smeaton's father, a lawyer, had been somewhat bewildered at his young son's strange passion for making tools and instruments. He considered that John, earmarked to follow in his legal footsteps, should be above such activities. A whiff of the common artisan hung about such pursuits. One of Smeaton's early endeavors, having watched a steam pump being installed in a nearby coal mine, was to make a smaller version. The young engineer then tested his creation on one of the fish ponds in front of the family house. The pump worked so well that it pumped the pond dry, the result being scores of dead fish and a very angry father.

Smeaton left school at sixteen to work in his father's office and train as an attorney. He brought with him, for a budding lawyer, some strange abilities. He could forge iron and steel, melt metals. In his small workshop he had tools (the lathes for wood and metal being homemade) for working in wood, ivory, and metals. His friends were always receiving beautifully worked presents.

In an attempt to wean him away from the workbench Smeaton's father, in 1742, sent him away to London where he hoped legal papers, ink, and quill pens would replace chisels and lathes. Vain hope. Smeaton wrote to his father pleading that the "bent of his genius" lay in worlds other than the law. A worried father, knowing his son, bowed to the inevitable and gave his blessing with a promise of financial assistance. Smeaton promptly went to work for a mathematical instrument maker.

Knight and Smeaton happened to share the same social orbits. Both were old-boys from Leeds Free Grammar School, and had lodgings close to each other, Knight's in Lincoln's Inn Fields and Smeaton's in Great Turnstile, a passage leading to the east side of Lincoln's Inn Fields.[1] Both attended meetings of the nearby Royal Society where Knight, a fully fledged F.R.S., Copley Medal recipient, and member of the Royal Society's exclusive dining club—in short, a big fish in London's natural philosopher's pond—acted as Smeaton's mentor. The gifted young instrument maker, in turn, was an ideal choice to make Knight's first compass.

A year following Knight's inspection of the *Dover's* compasses the two men unveiled their two new instruments before the assembled Fellows of the Royal Society. Knight, with his long, thin fingers poking out from his ruffled sleeves, pointed out its superior qualities compared to other marine compasses. And, indeed, it was an exquisitely made instrument. From its neatly dovetailed gleaming mahogany box holding the gimbals, to the brass-bowled compass with its black-and-white card decorated with a fleur-de-lis, it spoke elegant precision. Knight demonstrated the hinged glass top

---

[1]This area of London, until its enclosure with railings in 1735, had been a highly dangerous and filthy open field. The Act for its enclosure describes it as full of "vagabonds, common beggars and other persons [who] resort therein, whereby many robberies, assaults, outrages and enormities have been and continually are committed." It is now home to lawyers, solicitors, and barristers.

(no crudely puttied top on this instrument) giving easy access to card, compass needle, and pivot pin. Not that the needle would require constant remagnetizing (as was usual with soft-iron needles), for this rectangular section needle had been made from properly tempered steel and magnetized on Knight's machine. And the pivot pin in its artful mounting could be easily replaced when worn.

As an added bonus, Smeaton then demonstrated an azimuth compass he had fashioned using the new compass as a starting point. The word *azimuth* comes from the Arabic *as sumat*, meaning way or direction. In navigational terms, it is used to indicate the arc of the horizon, measured in degrees, between the observer's meridian and a vertical circle passing through any celestial body. An azimuth compass can also take bearings on headlands, capes, and prominent landmarks for position fixing. (Any ordinary compass will also serve the same purpose, although perhaps not to the accuracy of an azimuth compass. The method is simple. Bend down behind the binnacle and sight the target just above the compass. Close an eye and then chop a vertical stroke downward toward the compass, as if you were a martial arts expert hoping to break a brick. Follow this line with your eye and estimate where it crosses the compass card. Read off the bearing. You will have joined good company. In 1800, a master of HMS *Foudroyant* was known as "Chop the Binnacle" for this manner of taking a bearing. This method has served for generations of seamen, fishermen, and yachtsmen.)

Azimuth compasses were used to measure magnetic variation and had been in use since the middle of the seventeenth century. All worked on the same principle of comparing the sun's true azimuth as shown by its shadow on the compass

card, with the direction of the compass needle. Smeaton's
design used two slit vertical stiles, rather like gun sights, a horizontal index bar, cat-gut line, and a brass ring rimming the
compass card divided into half-degrees.

The demonstration before the Royal Society and the
description of the two compasses in the prestigious *Philosophical Transactions* constituted the opening shot in Knight's marketing campaign. So far he had borne all the costs of design and
manufacture. It was now time to start selling his compass, with
the Admiralty lining up in his sights as the obvious target.

Chapter 10

THE SHOCKS OF
TEMPESTUOUS SEAS

Commodore George Anson sailed from England
in 1740 with a fleet of six ships and over 1,900
men. The objective, in the words of his Admiralty
instructions, was to annoy and distress the Spanish in the
Pacific and, if possible, to capture one of the Spanish treasure
ships that sailed yearly from Acapulco in Mexico to Manila in
the Philippines.

He returned four years later with one ship, the *Centurion*,
and less than six hundred men. On the plus side was a treasure
valued at over £500,000, which made Anson, with his share of
the prize money, a very wealthy man. The treasure, carried in
thirty-two wagons escorted by the *Centurion*'s grinning seamen and more serious-looking officers, also provided a most
entertaining spectacle for Londoners as it was cheered
through Piccadilly, St. James's Street, Pall Mall, and then

through the City to the Tower. The whole parade was led by kettle drums, trumpets and French horns, followed by the first treasure wagon flying the English colors over the Spanish.

In 1747 Anson, now Admiral George Anson, made himself even wealthier by defeating a French squadron escorting a convoy bound for Canada. This treasure, worth over £300,000, was also paraded through the streets. Shrewd governments of the day knew the value of capitalizing on victories. In 1748 the authorized version of Anson's *Centurion* voyage was published and became an instant best seller. Its introduction stressed the importance of accurate charts and the collection of geographic and hydrographic information on all future voyages. Thus, through navigation, "Commerce and national interest may be greatly promoted."

When Knight unveiled his new compass, hoping to spread the gospel of its superiority, the portly and florid-faced Anson, a Fellow of the Royal Society, happened to be a man of vast influence at the Admiralty and a Commissioner on the Board of Longitude. Four other Fellows of the Royal Society, including Knight's friend, the President, Martin Folkes, also happened to be Commissioners. Knight was on the inside track to realizing his marketing dream.

A year after Knight's compass demonstration, Anson was appointed First Lord of the Admiralty and, much to the vexation of those comfortably burrowed in the naval bureaucracy, set about reforming the notoriously corrupt naval dockyards. He also visited Knight's lodgings in Lincoln's Inn Fields to inspect the doctor's magnetic machine, magnet bars, and compasses. Here he watched, like Folkes, Knight's masterly and theatrical demonstrations.

Events after Anson's visit moved with remarkable speed: the Navy Board was instructed to make an official examination of the new compass and magnetic bars; a few weeks later the Navy Board and Trinity House reported their findings. The same day of the report, April 4, 1751, compasses and magnet bars were ordered to be supplied to the forty-four-gun *Glory*, Captain Richard Howe (later Admiral Lord Howe), bound for the Guinea Coast; the forty-gun *Rainbow*, Captain George Rodney (later Admiral Lord Rodney), bound for Newfoundland; the sloop *Swan*, Captain John Jermy, bound for Barbados; and the sloops *Vulture*, Captain Thomas Wyatt, and *Fortune*, Captain Alexander Campbell, both sailing in the home waters of the English Channel. Howe, Rodney, and Jermy were also "required and directed to take an opportunity of seeing the said Doctor before you sail in order to your being thoroughly instructed by him in the use of his Magnetical Bars to prevent any mistakes and also informed of the improvements made on the compass."

Wyatt of the *Vulture*, a well-named vessel for such duties, was ordered to swoop "against the smugglers between the Downs and Beachy Head." The Admiralty also had other work for the *Vultures*: "We send you herewith information of some shipwrights being inveigled out of England and now at Boulogne, from whence they are to proceed abroad, and we do hereby require and direct you to have an eye upon any ship or vessel going from Boulogne and if you meet with any such persons in her as you have reason to suspect may be going on such a design, to seize them and bring them to England." Britain, with the world's largest merchant fleet, could ill afford to have such essential artisans as shipwrights working for potential enemies.

KNIGHT HAD NEVER tested his compass at sea before its presentation at Crane Court. And Smeaton, busy in his workshop, was embroiled in making other nautical instruments: an artificial horizon for taking celestial sights when the horizon was obscured and a ship's log for measuring speed and distance run. The latter consisted of a thin, twisted, oval brass plate 10 inches long and 2½ inches wide attached to a thin cord. When streamed astern of a vessel the twisted plate rotated, the rotations being counted by a series of dials. It was an idea that is still used. Smeaton's first attempt to calibrate this instrument was on the River Serpentine in Hyde Park. Here, in 1751, a stroller could have watched Smeaton in a rowing boat, trailing his creation like a fisherman trolling for fish.

A few weeks after these experiments both Smeaton and Knight were testing their creations aboard a small vessel in the Thames estuary. With them, according to Smeaton, was a "Mr William Hutchinson, an experienced seaman and master of a considerable merchant-ship." Hutchinson was indeed experienced. He had started his sailor's life as a young boy aboard a small collier, where he spent his time as a "cook, cabin-boy, and beer-drawer for the men." He had sailed to China aboard East Indiamen; fought in the Mediterranean as a privateer; been shipwrecked and cast adrift in a small boat. Without food the boat's crew had cast lots to see who would make a meal for the others. Hutchinson had been the unlucky man but was fortunately saved by a vessel's appearing over the horizon; and from then on he had kept that date as an anniversary day of "strict devotion."

This first compass test was followed by a most important one. On September 24, 1751, from the deck of His Majesty's sloop-of-war *Fortune*, anchored off Harwich, Knight and Smeaton were contemplating the gray waters of the North Sea, a sea dotted with the sails of hoys, coasting sloops, cod smacks, an endless stream of colliers, and the Harwich packet boat bound for Holland. Below this innocent and sail-dotted sea lay sandbanks that breached the surface at low tide like the backs of huge whales: the Long Sand, Gunfleet, Shipwash, Sunk, Kentish Knock. Pestering the entrance to Harwich lay even more ominous shoals upon which ships and sailors were sacrificed: Altar, Bone, Gristle, Glutton. This was a sea where lead line and an accurate compass were essential.

The next few days were frustrating ones. Bedeviled by light winds and calms the *Fortune* spent more hours at anchor than testing compasses and Smeaton's log line. By the end of the month the sloop had struggled north and come to anchor in the River Humber. Campbell's opinion of Knight's compass was cautiously optimistic: "Found one of the compasses to answer much better in light winds and with what sea we have had than the old compass."

Other opinions were not so cautious. Howe thought the compass preferable to the other compasses "except in stormy weather." Rodney's opinion was brutally frank: "Impossible to steer by."

Knight, unknowingly, had committed a serious fault in compass design. A compass card's weight should be equally balanced. Or, to put it into terms reeking of the schoolroom, its moment of inertia should be equal about all its diameters. In other words, when tilted from the horizontal it should not turn on its pivot. But that single needle ensured that it did turn. Such a compass, on a ship heading north and rolling in a beam sea coming from the east, would end up with the compass needle swinging east-west, and then swinging back to north-south on the top of the roll. Ironically, those lozenge-shaped needles so anathematized by Knight made a far steadier compass.

Knight's compass might have been flawed, but his marketing technique and influence with the Admiralty remained flawless. A year after the test aboard the *Fortune* the Board of Longitude, headed by Anson, awarded Knight £300. Later that year, all Royal Navy ships bound abroad were supplied with the doctor's compass and magnet bar.

During the summer of 1757, any of those outward-bound ships might have taken a bearing with their Smeaton-designed azimuth compass on the Eddystone Rocks lying south of Plymouth. The first Eddystone Lighthouse had been washed away in the great storm of 1703. The second was destroyed by fire in 1755. A third was now rising on top of the wave-swept rocks. This one happened to be designed by Smeaton and it was rising in a shape, and constructed of ingeniously dovetailed stone, that was to become the exemplar for all wave-buffeted lighthouses that followed. Smeaton was no longer making compasses; he had transformed himself into one of the new breed of civil engineers.

KNIGHT NOW HAD his compass and magnet bars made by his sole agent, George Adams, an instrument maker in Fleet Street

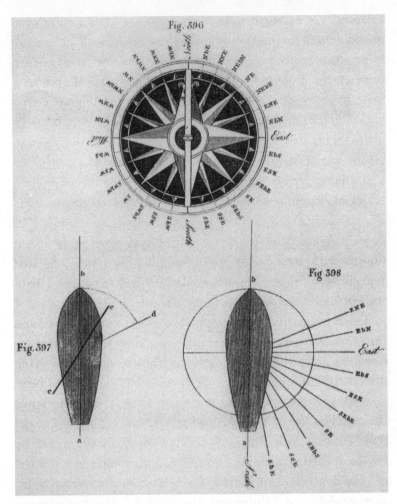

Illustration from *The Young Sea Officer's Sheet Anchor* by Darcy Lever (1808). The compass card, with the needle on top, is one from a Dr. Gowin Knight compass.

with a shop at the sign of Tycho Brahe's Head. This was no fly-by-night operation, but one that supplied mathematical instruments to royalty and had a worldwide clientele through a mail-order catalog. A Knight compass, selling at 45 shillings, was ten times more expensive than a compass with a lozenge-shaped needle. And they came with a marketing ploy copied from England's famous clock makers: a signed certificate from Dr. Gowin Knight F.R.S.: "I hereby certify that I have examined this compass and that it appears to be rightly constructed."

Testimonials to his compass, usually from Fellows of the Royal Society, appeared in books and pamphlets. William Hutchinson, the same Hutchinson who had helped test the first compass, had little time for those who muttered about its high price: "Where there are so many lives, and so much property depending on good compasses, I have been surprised and vexed to hear some people begrudge the price of Dr. *Knight's* improved steering and azimuth compasses, which I thought, when I bought one of each, not only deserved the price, but the inventor the thanks of the public as a trading nation and a maritime power for so great an improvement in that important instrument."

William Mountaine and James Dodson, on publishing an updated version of Halley's variation chart, wrote that compasses had been neglected: "Until of late years the judicious Dr. Gowin Knight, F.R.S., examined into their fabric and construction, employ'd his Magnetic Knowledge towards their Improvement, and has now reduced them to a considerable degree of Perfection."

John Robertson, a mathematician, librarian at the Royal Society, and author of *The Elements of Navigation*, wrote in his book that the faults in the compass and "several other imperfections have been happily removed by the labours of the truly celebrated Dr. *Gowin Knight*, F.R.S."

But, despite the puffery, the reports filtering back convinced Knight to set about improving his design to make it steadier in a seaway. Alas, having placed his trust in a single needle designed on magnetic principles, and knowing nothing of the compass card's dynamics at sea, his new design for the card's pivot socket only made matters worse. No matter; Knight was convinced that his new design was a vast improvement, and in 1766 it entered history as the first patented compass: "A New Method of Constructing Compasses in General Use So as to Prevent Them Being Affected by the Motion of the Ship."

DURING AUGUST OF 1766 a shore-side loafer might have speculated on the unusual amount of activity aboard three ships anchored off the Nore: the frigate *Dolphin*, the sloop-of-war *Swallow*, and the store-ship *Prince Frederick*. Rumor, according to the *Dolphin*'s master George Robertson, had it that the ships were to sail south to find a new continent. And Robertson, waxing lyrical on all the navigation instruments dear to his heart, noted: "We have received two New Invented Compasses . . . as Boath is Invented by the famos Doctor Knight I make no doubt of their answering what they are Intended for."

A week after sailing, Robertson was revising his opinion: "This Compass appears to be well constructed and I dare say will answer well in Smooth Water or on Land, but am Afraid the card is too heavy loaded, which in my opinion will make it traverse too quick in a Sea, when the Ship Roals or pitches much."

By the middle of December, sailing through Magellan's

Strait and watching the deerlike guanacos being hunted by the Patagonians, Robertson's opinion was confirmed: "The New Compass is a very fine instrument for observing the Variation when the Ships in smooth water. It is likeways very good for Observing the Variation ashoar, but will not answer at Sea in bad weather when the Ship has a quick motion. It then runs round and neaver Stands Steedy."

Six months later the Dolphins became the first Europeans to sight Tahiti and sample the delights of the ardent native girls with their flower-crowned heads and tattoed thighs. The delights were paid for with iron nails wrenched from the *Dolphin*'s hull and deck.

As Robertson surveyed Tahiti's shores, halfway across the world another Royal Navy master was likewise surveying the shores of a slightly more inhospitable island. The summer of 1767 was the fifth summer spent by Mr. James Cook, commander of the *Grenville*, surveying Newfoundland's indented coastline. Small boats were essential for Cook's meticulous survey and, over those five summers, Cook had found Knight's compass very unsteady compared to the old style, lozenge-needled, wooden-bowl compasses of his youth. In a letter to the Navy Board, thinking he would spend a sixth summer in Newfoundland, he made a request for those old-style compasses: "Doctor Knight's Stearing Compass's from their very quick motion are found to be very little use on Board small vessels at sea and I am very often oblig'd to have them in boats where . . . they are generally useless."

A month after writing that letter Cook was appointed commander of the *Endeavour* for a voyage into the Pacific. At two meetings of the Royal Society's Council (where he was introduced as "a proper person to be one of the Observers of

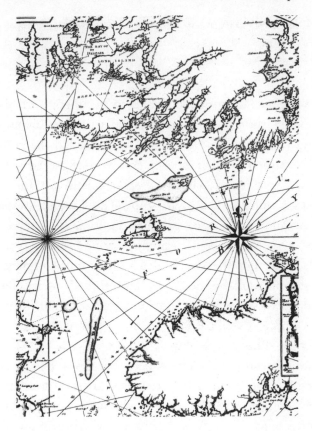

Portion of Newfoundland chart by James Cook, published in 1767. The chart shows the usual loxodromic lines and a compass rose giving the magnetic variation. This was the extraordinary quality of Cook's survey work. Cook had no great opinion of Knight's compasses, used on this survey, and thought them "generally useless" in small boats due to their quick motion.

the Transit of Venus"), he met Knight, a Council member. It was, perhaps, at these Council meetings that Knight persuaded Cook to revise his poor opinion of the doctor's compass. The newly patented compass, claimed Knight, was a vast improvement.

THE SUMMER OF 1768 happened to be uncomfortably hot, made even more wearisome by the turbulent behavior of striking coal heavers—the men who unloaded the colliers— and sympathetic collier sailors. But working-class solidarity collapsed into ruins when the sailors, held up for weeks, started to unload their own ships. The London waterfront soon became a war zone between fighting coal heavers and sailors. Other merchant seamen, infected by the anarchy, paraded through the streets demanding more money. They boarded outward-bound ships and forced their crews to unbend sails, strike yards and topmasts, and then forced the crews ashore. Even hatters—in an age when every person wore a hat—went on strike.

It was in this uneasy atmosphere that Cook had to prepare his vessel. One of his many requests to the Admiralty concerned compasses: "Doctor Knight hath got an Azimuth Compass of an Improv'd construction which may prove to be of more general use than the old ones: please to move my Lords Commissioners of the Admiralty to order the Endeavour Bark under my command to be supplyed with it."

Smeaton, his Eddystone Lighthouse having made his reputation, was also called upon, this time to design a portable wood-and-canvas observatory to house the Transit of Venus instruments.

Cook and the *Endeavour* sailed from Plymouth, Smeaton's Eddystone Lighthouse pricking the horizon, on August 25, 1768. Three years later, after the Transit of Venus observations in Tahiti, the charting of New Zealand, near fatal shipwreck in Australia, and death and disease in Batavia, the weather-beaten *Endeavour* came to anchor in the Downs. Leaving his anchored ship, Cook sped to London by rattling coach to report to the Admiralty. With him went all the reports on the voyage: among them was one on Knight's compass: "Agreeable to their Lordships Commands I am to acquaint you that I never once was able to make use of the Compass in a troubled Sea ..."

An equally damning assessment, but a more public one, had appeared while Cook was in the Pacific. In William Falconer's *An Universal Dictionary of the Marine*, under the entry for Compass, appeared this footnote:

> It is necessary to observe here that the principal, and indeed the only circumstance in which Knight's compasses are superior to those which have hitherto obtained, is, that their needles being tempered much higher than the usual, are thereby enabled to contain a much greater quantity of the magnetical stream, which is certainly a real advantage. But, on the other hand, experience sufficiently proves, and truth obliges us to remark, that the methods he has taken to balance the card with more accuracy than has been formerly attempted, have rendered it by far too delicate to encounter the shocks of tempestuous seas.

KNIGHT, A DECADE before this blast from Falconer, had been appointed as the first Principal Librarian, or Director, of the newly formed British Museum. It was Knight who drew up the plan for the distribution of Sir Hans Sloane's natural history collection. Hidden away in the basement went "Monsters and anatomical preparations . . . with the skeletons, and other parts of anatomy . . . as all these are not proper objects for all persons, particularly women with child."

The British Museum was opened to the public in 1759. But it took tenacity to break through the thicket of regulations if you wanted to see the exhibits. A visitor had first to make an application at least one day before the visit. A second visit was then required to receive the admission ticket, which, displaying superb bureaucratic cunning, could not be used that day. Thus, it required three journeys to the museum before you could see the offerings. All visitors (limited to sixty people a day) were then ushered around in a single group. The tour ended on the ground floor where Knight displayed his magnetic machine.

Knight and his three assistants acted as librarians, catalogued the books and collections, acted as guides and watchmen. The assistants were not free to leave the building at night without leave from Knight. Within a few months they were all feuding. Thomas Gray, the poet, summed up the Museum's lowering atmosphere in a letter: "I often pass four hours a day in the stillness & solitude of the reading room. . . . When I call it peacefull, you are to understand it only of us Visiters, for the Society itself, Trustees, & all, are up in arms, like the Fellows of a College. The Keepers have broke off all intercourse with one another. . . . Dr Knight has wall'd up the passage to the little-House, because some of the rest were obliged to pass by one of his windows on the way to it."

The acerbic Knight, regretted by few, died in the British Museum. Six years later, blithely swimming against the tide of opinion flowing from their seagoing officers, the Admiralty ordered that Knight's compass become the official compass for all of His Majesty's ships.

One ship issued with a Knight's compass was Bligh's *Bounty*. The mutineers removed the compass before the *Bounty* was scuttled off Pitcairn Island. In 1808 the mutineers' hideaway was discovered by an American sealer, Captain Mayhew Folger of the *Topaz*, and the last surviving mutineer gave Folger the compass. In 1813, Folger sent it to the admiral commanding the Royal Navy's North American Station:

> This compass I put in repair on board my ship, and made use of it on my homeward passage, since which a new card has been put to it by an instrument maker in Boston. I now forward it to your lordship, thinking there will be a kind of satisfaction in receiving it, merely from the extraordinary circumstances attending it.

The compass, however, having survived such adventures, has vanished, just as most of Knight's compasses have vanished. (Two reproductions of Knight's compass can be seen at Portsmouth in the binnacle of HMS *Victory*.) Nevertheless, the legacy of Knight's compass, so "ticklish and subject to vibrate," would inspire other compass makers to grapple with the problem.

But another problem was looming over the horizon for the sailor and his compass, a problem that Captain Matthew Flinders of the Royal Navy was to spend years of his life attempting to solve.

Chapter 11

ANY OLD IRON,
ANY OLD IRON

On a sparkling summer day of 1801, with a brisk easterly wind ruffling the blue of the English Channel into flecks of white, any interested coast-watcher with a spyglass, lying on the close-cropped grass at the edge of Devon's Start Point and pondering upon the chances of an invasion from France, would have seen a bluff-bowed, three-masted vessel, courses, top-gallants, and topsails curved in the fresh breeze, white water foaming at those bluff bows, making her way down Channel and out into the Atlantic.

Start Point, in its turn, was also proving of worrying interest to observers on the rolling vessel, two sets of compass bearings, one for measuring magnetic variation and the other on Start Point, being the cause. The bearings had been taken at two different positions on the ship, one at the binnacle, just forward of the mizenmast, and the other on the booms, just forward of the mainmast. The readings disagreed by 5 degrees. The same had happened some 65 miles astern when off St. Alban's Head.

This was no "chop the binnacle" measurement, but one taken with a new type of azimuth compass designed by a certain Mr. Ralph Walker and made by George Adams of Fleet Street, which gave the compass a near impeccable pedigree.

RALPH WALKER, UNLIKE Dr. Gowin Knight, had spent years at sea in the merchant service, working his way from seaman to his own command. In 1783 he had settled as a planter in Jamaica. But those years at sea had set him to thinking, like William Whiston, on finding the longitude at sea by compass variation—and thus collecting a financial reward, as had Knight, from the Board of Longitude.

For the next ten years, Walker worked at improving the machinery used on the sugar plantations and, in the cool of the evenings, at a method for finding the longitude. It was a complicated method and based, like all the theoretical methods of finding the longitude by magnetic variation, on the idea that the lines of magnetic variation were spaced with regular and military precision around the world: a theory at odds with Halley's World Chart of Magnetic Variation. No matter; Walker attributed these annoying anomalies to deflection by land masses. Walker's industry eventually produced a book of tables giving the variation for every 2 degrees of latitude and 1 of longitude. This theoretical hurdle crossed, and knowing the uncertainty of measuring magnetic variation at sea, he set about designing and making a new type of azimuth compass which could be used at any time when the sun was above the horizon.

His design was elegant and clever. Above the compass bowl he fitted his own version of a universal sundial. This could be adjusted for latitude and the sun's declination. The compass was then turned so that a slider with a small aperture threw a spot of sunlight on the center of a semicircular scale. The navigator then noted the number of degrees between the lubber's line marked on the compass bowl and magnetic north on the compass card. This reading gave the magnetic variation. The navigator then entered Walker's book of tables with the vessel's latitude and magnetic variation to find the longitude. That was the theory conceived in the soft and gentle warmth of Jamaican evenings and its lush landscapes. But it needed testing.

In the June of 1793 a convoy of merchant ships, shepherded by the Royal Navy, sailed from Jamaica bound for England. One of the naval vessels, HMS *Providence*, carried a rather bizarre cargo. Captain William Bligh, his voyage in the *Bounty* having been cut short by the famous mutiny, had completed the *Bounty*'s original objective of delivering thousands of Tahitian breadfruit plants to the West Indies. But the *Providence* still carried some seven hundred plants of different species, all potted up in tubs, destined for the Royal Gardens at Kew. Bligh could congratulate himself on a most satisfactory voyage. And on the even more satisfactory thousand guineas presented to him by the Jamaican Assembly for delivering the breadfruit plants. The breadfruit, it was hoped, being much prized as a food by the Tahitians, would also be prized by the slaves working on the sugar plantations.

Also on board, traveling as a passenger at the request of

Jamaica's governor, was Ralph Walker, chancing his all to lay his new compass and his theory before the Admiralty.

Two months later, both plants and Walker were safe ashore in England, as was a young midshipman from the *Providence*, Matthew Flinders, who made his way to London's Soho Square to deliver a letter for Sir Joseph Banks, President of the Royal Society. The letter was from James Wiles, one of the *Providence*'s botanists and a Banks appointee for the breadfruit voyage. Wiles had remained in the West Indies to nurture the transported plants. The threads making up Flinders' future life were being woven around him.

Walker, in a whirl of activity, presented his theory and compass to the Board of Longitude, arranged for the publication of his longitude method, and, more importantly, persuaded the Admiralty to have his compass tested at sea aboard four of His Majesty's ships.[1]

Nevil Maskelyne, the Astronomer Royal, in a report to the Board of Longitude, was less than enthusiastic about Walker's theory. But navigators embraced the new compass (if not the longitude theory), for the compass proved far steadier than Knight's compass in a seaway, and far superior for measuring magnetic variation. In October 1795, it was adopted by the Royal Navy. But, as the compass was more expensive than inferior makes, it was only issued to ships on a penny-pinching precondition: captains and masters had to produce a certificate showing they were proficient in its use.

But before this happy event for Walker, a report on his compass came from Murdo Downie, master of HMS *Glory*,

---

[1]They were the *Invincible*, *Glory*, *Queen*, and *Lynx*. Walker sailed aboard the *Invincible*.

who had noticed the discrepancy in magnetic variation when measured in different positions aboard the *Glory*:

> I am pretty well convinced that the quantity and vicinity of iron in most ships has an effect in attracting the needle; for it is found by experience that the needle will not always point in the same direction when placed in different parts of the ship; also it is rarely found that two ships steering the same course by their respective compasses will go exactly parallel to each other; yet these compasses, when compared on board the same ship, will agree exactly.

Downie's observations were not new. In 1538, João de Castro, the Chief Pilot of the Portuguese India Fleet, had noted the puzzling behavior of a compass needle and finally narrowed it down to an iron cannon. Observations for magnetic variation, Castro surmised, were "troubled by the proximity of artillery pieces, anchors and other iron." It is the first mention in print of the effect of iron on a ship's compass. A century later, Captain John Smith (sometime Governor of Virginia and Admiral of New England, as he styled himself on the title page of his 1627 *Sea Grammar*) warned against having iron nails in the compass box. Sixty years later, Captain Samuel Sturmy, in his *Mariners' Magazine*, was warning that the compass needle could be drawn aside by guns in the steerage or any iron near the compass.

Captain James Cook was well aware of the dangers of iron near the compass. In 1776, aboard the *Resolution*, he entered an acid comment into his journal on "the advocates for Variation" who hoped to find their longitude by such a method: "Whoever imagines he can find the Variation within a degree will very often find himself much deceived, for besides the imperfection which may be in the construction of the Instrument or power of the Needle, the Motion of the Ship, or attraction of Iron work, or some other cause not yet discovered, will frequently cause far greater errors than this."[2]

The amiable William Wales, astronomer on Cook's second circumnavigation, whose instructions from the Board of Longitude included measuring the magnetic variation and dip, also noted many puzzling matters when it came to the compass. Wales noted that a change in the ship's course was enough to alter the measured magnetic variation from 3 degrees to a staggering 10 degrees. From these and other observations Wales came to the conclusion that "variations observed with a ship's head in different positions and even at different parts of her will vary materially from one another."

This troubling effect is known as compass deviation. And Matthew Flinders, the young man off the *Providence*, was to spend years of his life investigating this compass mystery. For this was a magnetic Siren song which could lure sailors to shipwreck and death.

---

[2]Certain eminent authorities have claimed that Cook (who they said kept keys to leg irons in the binnacle) and Bligh (who they said kept pistols in the binnacle) were unaware that iron affected the compass. But the quotation shows that Cook was well aware of this. The keys in the binnacle incident (October 30, 1777) occurred with the *Resolution* at anchor in Tahiti. The keys, stowed in the binnacle, were for leg irons used on a native who, much to Cook's fury, had stolen a precious sextant. The Bligh pistol incident also happened at Tahiti. After three seamen had deserted from the *Bounty*, pistols were stowed in the binnacle but were soon removed to the master's cabin. These were the pistols used by Fletcher Christian in the mutiny. However, the *Resolution* and *Bounty* never sailed with keys or pistols close to the compass.

.   .   .

THE COMPASS BEARINGS taken off the appositely named Start Point in 1801 were the start of Flinders' quest—a quest which was to turn into an odyssey of Homeric proportions. An odyssey that encompassed dangers and discomforts, disease, shipmates drowned, shipwreck, a voyage in an open boat, and years of captivity on an alien island before his return to a wait-ing wife. Luckily for the dark-haired Flinders, bracing himself against the ship's roll as he watched the English coast vanish astern, a scar over his left eye giving him a slightly piratical look, all these sorrows were yet shrouded by the future.

It is perhaps time to take a look at this twenty-seven-year-old officer. And in looking it is impossible not to compare him with James Cook, Flinders' idol. At twenty-seven, Cook was still a mate aboard a collier sailing through the North Sea's treacherous shoals and sandbanks, although, it has to be admitted, only months away from joining the Royal Navy as an able seaman. Thirteen years later, Cook was sailing down Channel with his own command at the start of one of the world's great voyages of exploration: the *Endeavour* circum-navigation of 1768–1771. The *Endeavour*, like Flinders' com-mand, the *Investigator*, had been built as a collier for the bringing of coals from England's northern counties to Lon-don's hearths. Colliers were not sleek or fast. But, being beamy, shoal-draft, flat-bottomed, and with large stowage capacity for months of provisions and stores, they made ideal exploration vessels.

The *Investigator*, just like Cook's *Endeavour*, carried live-stock destined for the table, dogs for hunting, cats for rat catching, and an assorted party of civilians: botanists, artists, and astronomers. The two ships might have been mirror images. But the mirror image of their commanders is distorted by the difference in their characters.

Both were superb seamen and navigators. Both paid particular attention to the care and health of their men. Both were passionate men when it came to the craft of surveying: but there was a febrile, a feverish quality about Flinders, not apparent in the pragmatic, calm, rock-solid Cook. Cook's officers might wryly note that "the old boy has been tipping a *heiva*"[3] when they had fallen from grace in some matter of seamanship or survey, but these were mere line squalls, devastating but soon over. Flinders brooded. He belonged to the Romantic Movement filtering through from Germany, the cultural and literary movement epitomized by Schiller, Goethe, Rousseau, Coleridge, Keats, Wordsworth, Shelley, Byron, Scott.

Flinders, born in the flat landscapes of Lincolnshire, came from three generations of surgeons. And his father intended that his son should follow in the family tradition. But Matthew had joined the Royal Navy at the age of sixteen. Years later, in answering the biographical question from the *Naval Chronicle*: "Juvenile or miscellaneous anecdotes illustrative of individual character?" he wrote: "Induced to go to sea against the wish of friends, from reading *Robinson Crusoe*."

The eleven years following the joining of his first ship were packed with storms and calms, sea battles, strange sights, the making of firm friendships. But it was the *Providence* voyage to Tahiti with William Bligh that gave him the taste for

---

[3]One of Cook's rueful midshipman, at the receiving end of a *heiva*, defined it as follows: "*Heiva*, the name of the dances of the Southern Islanders, which bore so great a resemblance to the violent motions and stampings on the Deck of Capt. Cook in the paroxysms of passion."

exploration and surveying. And Bligh, having sailed with James Cook, was a direct link to the man whom Flinders worshiped.

Flinders, two years after arriving home aboard the *Providence*, was back in the southern hemisphere, this time as the master's mate aboard HMS *Reliance*, carrying Captain John Hunter to the seven-year-old convict settlement at Sydney Cove in New South Wales. Hunter, the new governor, was a very capable surveyor and he had requested from the Admiralty a most comprehensive collection of navigational and surveying instruments, charts, drawing paper, and drafting tools. The list included a Walker azimuth compass. But the Admiralty dug in its heels at the request for stationery and drafting instruments. "I am commanded by Their Lordships," came the brusque reply, "to acquaint you that as the sloops are not ordered to proceed on a voyage of discovery they do not think it necessary that they should be furnished with articles mentioned in the said list." Hunter, knowing the importance of survey work, asked their Lordships to reconsider. Back came a grudging reply: "Their Lordships have ordered you to be supplied with a small quantity of stationery for the sole purpose and use of such maritime surveys as you may occasionally see it necessary to employ the ship's boats upon." It was a small victory for Hunter, for the British were lamentably ignorant of the coastline close to their infant colony, and even more ignorant of the interior.

On the voyage to Sydney, Flinders struck up a close and lasting friendship with George Bass, the *Reliance*'s surgeon. The tall, debonair Bass, three years older than Flinders, happened to be another Lincolnshire man, and his sardonic humor did much to leaven the gravity of the more serious-

minded Flinders. Between them they hatched plans for exploration.

Their first expeditionary vessel verges on the ludicrous.
Seven weeks after sailing into Sydney's harbor, Port Jackson,
aboard a ship that sported three masts, fifteen sails, and sixteen
cannon, Flinders, Bass, and William Martin (Bass's boy servant)
were sailing out into the Pacific in an open boat which measured 8 feet along the keel. The aptly named *Tom Thumb* was
owned by Bass. Bobbing like a child's bathtub duck in the
Pacific swells, it gave an air of youthful exuberance to this
mini-expedition as Flinders steered, Bass tended sail, and
young Martin bailed. A week later, after exploring the head-
waters of Botany Bay, the trio were back at Sydney Cove and
reporting to Hunter. And the tiny *Tom Thumb* entered Australian history.

Over the next four years, sailing in a series of ramshackle
craft that would make a modern yachtsman flinch with horror
or applaud in admiration, they discovered the strait, Bass Strait,
which separates Tasmania from the mainland, and made the
first circumnavigation of the island. It was on one of these
many voyages that Flinders, now Lieutenant Flinders, sailing
aboard the 45-foot schooner *Francis*, noted puzzling changes
in the magnetic variation when the schooner's course was
altered. For the twenty-four-year-old lieutenant, as with William
Wales, the compass's antics were a puzzle. But Flinders, unlike
the astronomer, was to crack the puzzle.

IN THE AUTUMN of 1800 a letter and package arrived at 32
Soho Square, London residence of Sir Joseph Banks, President
of the Royal Society. Behind the house's imposing facade lay

reception rooms, an Adam drawing room, a huge library, a herbarium, and numerous glass-fronted cabinets holding Sir Joseph's collections. All these surroundings (for this was a hospitable house with a welcoming host) were familiar to an astonishing range of people, irrespective of age, class, nationality. Here could be found army and navy surgeons, doctors, astronomers, mathematicians, physicists, geologists, gardeners, explorers, and whalers; all of them either sipping tea from china bowls or carving meat at the breakfast and dinner table.

Their host, no longer the lean young man who had sailed with James Cook aboard the *Endeavour* and tasted the delights of Tahiti and the weevils in the ship's biscuits, was now overweight and plagued by gout. But his influence was as large as his girth. King George III listened to him as did the men in government. He had turned the Royal Gardens at Kew into the world's premier botanical garden, and suggested the penal settlement in New South Wales, Bligh's voyages to collect breadfruit plants, Mungo Park's travels in Nigeria. The range of interests of this tireless man was all-encompassing. (Two Johnsonian quotes spring to mind when thinking of Banks. The first one, that the breadth and depth of a man's mind is in direct proportion to the extent of his curiosity: by this yardstick Banks romps home a winner. The other, his famous definition of a patron: "Commonly a wretch who supports with insolence, and is paid with flattery." On this one Banks, a most generous and kindly patron, would fail.)

The package arriving at Soho Square had been sent by Flinders, who had returned to England aboard the *Reliance*, which was now lying at Spithead after a seven-month voyage from Sydney. The package contained seeds and plants (always dear to Banks's heart) and a letter from Flinders. The letter, honed and perfected during the voyage, held Flinders' bid for the future. And it was an audacious bid: a proposal to survey the unknown coast of New Holland—today's Australia—with Flinders as commander of any such expedition.

It was hardly the best of times for an unknown lieutenant of only two years seniority to make such a proposal. Britain was fighting a war with France led by her aggressive First Consul, Napolean Bonaparte, who was gleefully rampaging across Europe and knocking down Britain's allies like ninepins. The Royal Navy might relish Nelson's defeat of the French at the Battle of the Nile, but it was still stretched to breaking point on blockade and convoy duties. And new storm warnings threatened from the north in the shape of the Armed Neutrality, engineered by Bonaparte, between Russia, Prussia, Denmark, and Sweden, which could cut off supplies of hemp, timber, and tar from the Baltic, products that were essential for the ships that made up the world's largest merchant and fighting navies.

It proved a long and anxious nine weeks for Flinders. His letter and seeds appeared to have fallen on barren ground. And then, magically, happily, he received a reply: Sir Joseph had been ill but would be delighted to see Flinders at Soho Square.

Flinders, in London to arrange the publication of his charts and sailing directions for Bass Strait and the New South Wales coastline, wasted no time in calling upon Banks. After this crucial meeting came a flurry of activity: the Admiralty ordered the Navy Board to slip the armed ship *Xenophon*, make good any defects, and supply her with provisions for six months to sail on foreign duties. All this suggests that Banks had mulled over Flinders' suggestions with his great friend

Earl Spencer, First Lord of the Admiralty, before the crucial meeting between Flinders and Banks at Soho Square.

The *Xenophon*, built as a collier, had been bought into the Royal Navy and used as an armed ship to escort convoys. Smaller than Cook's *Endeavour* and *Resolution*, she had been weakened by having gun ports cut in her hull to mount carronades. On January 19, 1801, her name was changed to His Majesty's Sloop *Investigator*. Within days, Flinders was aboard and mustering the crew on deck. And with a pinching cold wind and flurries of snow sweeping around their ears, Flinders read himself in as their commissioned commander. His gamble for his dream command, against all odds, had won through.

Chapter 12

THE BOOK OF BEARINGS

As the *Investigator* butted her way down Channel off Start Point and out into the Atlantic swells on that summer day in 1801, on the opposite side of the globe two French vessels were gingerly exploring New Holland's arid and desolate western coast. Ironically, it was the sailing of this French expedition in the autumn of 1800 that had helped Flinders achieve his ambitious exploration plans and speeded the preparations for the *Investigator*'s voyage.

Passports might seem unusual items for ships to carry, but in those more civilized days it was usual for nations at war to give passports, or safe-conduct passes, to their enemy's ships when those ships were sailing on voyages of scientific exploration. In late June of 1800, before Flinders had returned in the *Reliance*, the British government had given such passports to Captain Nicolas Baudin and his two ships, the *Géographe* and the *Naturaliste*. This expedition, claimed the suave French officials, was bound on a scientific voyage around the world. In truth, the main objective was the southwest coast of New Holland. Such intelligence, seeping across the Channel and into Whitehall, was enough to set alarm bells ringing at the Admiralty and behind the Ionic columns of the East India Company's building in the City.

Île de France (present-day Mauritius), an island in the Indian Ocean, happened to be a hornet's nest of French privateers preying on the Company's ships as they made their homeward-bound passage stuffed full of eastern riches. And the thought of the French setting up another hornet's nest on the shores of New Holland was enough to give the entire Board of Directors one massive, kindred, apoplectic fit. Plus— and this was worse—the French expedition had sailed with the full approval of Napoleon Bonaparte, who was no more to be trusted than a mad dog. And then, magically, Flinders had appeared from the wings with his expedition plans. It was possible that such an expedition might discover a shorter route to the East Indies and China. Moreover—and this appealed mightily to the British government and the East India Company—it could also keep a wary eye cocked on any French colonizing tricks in New Holland.

The *Investigator*, in her turn, carried a passport from the French. It had taken time to arrive from France, an annoying wait which had kept the *Investigator* swinging at anchor for many weeks. But this was no French trick, just pure incompetence and laziness by the British Foreign Office in applying for the passport. Banks, on the other hand, in organizing the *Investigator*'s scientific staff, had shown no laziness, only remarkable energy. Two of the most telling sentences in all of the correspondence generated during the birth of the *Investigator's* expedition are in a letter from Evans Nepean, Secretary of the Admiralty, to Banks: "Any proposal you make will be approved. The whole is left entirely to your decision."

THE *INVESTIGATOR*'S PASSAGE to New Holland's southwest coast took five months. And those five months revealed that she spewed her caulking in a seaway—at one point she was making five inches of water an hour through her leaking seams—and whose masts, spars, and rigging needed constant attention.

Of all the Australian coastlines surveyed by Flinders, the coast rimming the Great Australian Bight is the most inhospitable and terrible. Here, for hundreds of miles, stretch cliffs over 300 feet high with an upper layer of dark sandstone laid on limestone, the exposed face of an ancient seabed. Not a single good anchorage can be found along this coast. Rolling up from the Southern Ocean sweep massive swells which, rearing up on each other to make wave-trains of frightening height, burst with explosive force at the cliff's base. It's a country of extremes: by day unbearably hot, by night freezing cold. Seemingly eternal winds scour this primeval coast. The arid and desolate land that stretches north in all its aridity is known as the Nullarbor Plain—the treeless plain.

Forty years after Flinders surveyed this awful coast an extraordinarily tough young Englishman, Edward John Eyre, looking to pioneer an overland stock route from east to west, came close to death along its shores. Peering over the cliff's edge he saw at its base bleached bones of whales, shells of giant turtles, wreckage of long-lost ships. Eyre was the first

European to journey across this terrible region. Even the aborigines avoided it. For a map he used a Flinders chart.

The making of those charts, as with all marine survey work, had been repetitive, time-consuming, tedious, dangerous. The charts had been made at the cost of lives, with eight men drowned when their cutter was sunk in a tide rip. One of the men happened to be John Thistle, the *Investigator's* master, who had sailed with both Bass and Flinders in Australian waters.

The rank of master no longer exists in the Royal Navy, and even in Flinders' day the rank was reaching the end of its long, historic, and honorable life. A master's duties included the ship's navigation and the making of charts and sailing directions when sailing in strange waters. Unlike commissioned officers—lieutenants, captains, and the like—who bathed in the glow of a commission from the crown, masters were appointed by the Navy Board. Both Cook and Bligh had served as masters before receiving their commission as lieutenants.

Thistle's death came as a severe blow to Flinders. The two men had sailed thousands of miles together and forged strong bonds of friendship and mutual respect. Since that day in the English Channel when the difference in measurements of magnetic variation had been noted, both Flinders and Thistle had made careful note of all bearings and magnetic variations taken from different positions aboard the ship. Hidden iron was the obvious answer to any errors. And Flinders ordered that a strict search be made for "sail needles, marline spikes or other implements of iron which might have been left near the binnacle." Two 12-pound carronades close to the binnacle were removed and struck down into the hold. All bearings were now taken from the binnacle, a position, thought Flinders, now free from any of the sly and hidden effects of iron—only to find another strange anomaly.

As the *Investigator* altered her heading—from east to west, say—so did the bearing on a prominent land feature. In an ideal world the bearing should have stayed the same. And these differences appeared to be related to the amount of magnetic dip: the larger the dip, the larger the error. And, adding even more confusion, the differences were reversed when in the southern or northern hemisphere. In the English Channel, for instance, the compass gave too much west variation when the ship was heading west. In Australian waters, when steering west, the measured variation was always too little. Here was a compass riddle waiting to be solved.

THE ANSWER TO the riddle lies in the archives of the Hydrographic Office in Devon. It's known as the *Book of Bearings.* Or, to give this 329-page monument to Flinders' efforts its full title: *Bearings taken on board His Majesty's Ship Investigator whilst exploring the coast of Terra Australis, by Mattw. Flinders—Commr. 1801, 2, and 3.* This is Flinders' Philosopher's Stone. A few pounds of paper that he used to transmute thousands of compass bearings (bearings which also noted the date, time, ship's heading, magnetic dip, latitude, and longitude) into a rational explanation for the compass needle's strange behavior.

By the time Flinders had completed his survey of Australia's south coast, from Cape Leeuwin to Cape Howe, his *Book of Bearings* held close to a hundred pages packed with 2,500 compass bearings.

SIX WEEKS AFTER losing a tenth of the *Investigator's* crew in the loss of their cutter, a sail was sighted on the horizon.

Flinders cleared for action and hoisted the ensign. The approaching vessel hoisted a French ensign. Both vessels hove-to, Flinders ordered a boat to be lowered, and was then rowed across the choppy sea to the enemy vessel. She was the *Géographe*. On boarding the vessel Flinders exchanged passports with Baudin and they held a stilted conversation in English—Baudin insisted on speaking in his fractured English, even though Flinders had brought along the *Investigator*'s French-speaking botanist. Meanwhile, Flinders' boat's crew having aroused the envy of the French sailors with their hats made from kangaroo skin, happily chattered away in the lingua franca of the sea.

The *Naturaliste* had parted company from the *Géographe* in a gale, and the latter had lost eight men of a boat's crew who had gone astray.[1] The English and French parted next day. But not before Flinders, having swept a critical eye across the *Géographe*, had come to the conclusion that the French corvette was in slovenly shape and her crew showing the ominous signs of scurvy. The two ships were to meet again, but in very different circumstances.

The *Investigator*, three months after her meeting with the French, sailed into Sydney Cove and anchored off Government Wharf. Flinders congratulated himself on the health of his crew as "there was not a single individual on board who was not upon deck working the ship into harbour." The Investigators, after the flurry of anchoring and making a neat harbor-stow of the sails, gazed with pleasure and curiosity (they had not clapped eyes on anything like this for six months) at the neat-looking settlement with its whitewashed

[1] The men, in fact, had not been lost. They were rescued in the Bass Strait by an English sealer, the *Harrington*.

Government House and green garden sloping down toward the shore, the smaller cottages lined up with military precision within their picket fences, the dusty streets, the slow turning blades of the windmills, the red-uniformed soldiers—"lobster-backs," to the sailors—at the wharf and, far more interesting, the petticoated female figures. And the more knowledgeable among the hands pointed out to the more ignorant the low-life area known as The Rocks.

Of equal interest was the anchorage. Lying at anchor were two other Royal Navy vessels, the small surveying brig *Lady Nelson* and the *Porpoise*; an English whaler, the *Speedy*; a priva-teer brig, the *Margaret*; and the hulk *Supply*. But by far the largest vessel was the French *Naturaliste*.

Captain Jacques Hamelin, with his ship running short of provisions and some of the crew laid low with scurvy, had decided to make use of his passport and seek help from the English at Sydney. Hamelin had arrived two weeks before Flinders and had been astounded at his welcome. His sick sea-men had been taken ashore into the hospital, provisions sup-plied (even though the colony was on short rations), and Governor Philip Gidley King had thrown a supper and ball for the French officers. Passports, it would appear, were useful items.

A month later the *Géographe* was given a similar welcome. But this time, so scurvy-ridden were the crew, the ship had to be helped by men and boats from the colony—including a boat from the *Investigator*—and towed into the anchorage. King sent baskets of fresh vegetables from his garden and opened the hospital for the sick French. And, since the arrival of the *Naturaliste*, he had the pleasure of informing Baudin that peace had been signed between their two countries.

It takes an impossible leap of the imagination, standing at Bennelong Point today and looking across Sydney Cove to Dawes Point, to picture the same view during those few weeks of 1802 when the British and French seamen were preparing for another bout of surveying. Today one stands in the shadow of the Sydney Opera House, with its billowing, spinnaker-like roofs, and looks across to the southern end of Sydney Harbour Bridge. Ferry boats shuttle back and forth from Circular Quay. To the left rise graceless clumps of skyscrapers as if dumped from a child's cereal box of instant plastic cityscapes. At the head of the Cove sweeps the vandalism of the Cahill Expressway with its ceaseless flow of cars. Behind the Opera House, however, stretch the delights of the Royal Botanic Gardens, the legacy of Captain Arthur Phillip, the colony's first governor, who declared the area to the east of Government House remain forever in the public domain.

The Sydney Cove of 1802, by contrast, would have presented a busy but more harmonious, almost Arcadian, scene— a handful of vessels swinging at anchor, small fishing boats slipping by under sail, aborigines paddling their canoes. On shore, close to where the present-day Opera House hoists those improbable roofs, the *Géographe* lay careened, her copper sheathing being repaired. Pitched close to the sandy shore were tents from the British and French ships. Two of the tents housed astronomical observatories, British and French. Armorers worked at forges, coopers on casks. The *Naturaliste* rode at anchor, having been fumigated for five days with a potent mixture of sulphur, gunpowder, and arsenic in an effort to kill the thousands of rats that had gnawed at sails and ropes, eaten wheat and rice—and had even had the gall to chew at

Hamelin's journal. Such were the scenes painted by artists from the two expeditions.

What the artists could not capture was Flinders bent over the *Book of Bearings* and working on his draft charts. Nor the four convict carpenters reducing the height of the bulwarks along the *Investigator*'s quarter-deck. The high bulwarks had been a great inconvenience when taking bearings from the binnacle. At times, with the ship heeled under a press of sail, Flinders had been forced to stand on the binnacle.

On July 14—Bastille Day—the waters of Sydney Cove became busy with cutters, gigs, and launches ferrying the French naval officers and the colony's officers and officials to the *Investigator*. Here, buttoned tight into their uniforms, they dined in the great cabin, exchanged stilted courtesies, toasted the health of the First Consul, toasted the health of King George III. This alcohol-fueled *entente cordiale* was to prove short-lived.

A week after that dinner party the *Investigator*, accompanied by a small consort, the *Lady Nelson*, sailed out into the Pacific bound on the second stage of her survey—New Holland's east coast and the Gulf of Carpentaria.

The *Lady Nelson*, for the connoisseur of the more bizarre aspects of naval architecture, is an absolute delight. She had been designed by Captain John Schanck of the Royal Navy and built at Deptford. She was a small vessel, measuring in at 60 tons and 52 feet overall. Her claim to fame rests with her three sliding keels (dagger boards), which gave her a draft of about 10 feet when they were down, and 6 feet when they were retracted. Schanck had originally designed her with a single mast and fore-and-aft cutter rig, but had been persuaded by Philip Gidley King (then in England before sailing to become governor of New South Wales) to change the rig to a two-masted brig. King claimed that "few seamen know anything about the management of a cutter, her being constructed into a brig would make her more manageable to the generality of seamen."The change proved disastrous. His Majesty's Armed Surveying Brig *Lady Nelson* proved a wretched vessel when trying to struggle to windward, all that she could usually manage being a most inglorious, crablike slide to leeward. No matter; this odd little vessel, this perverse but somehow likable monument to naval architecture, managed to sail from England to New South Wales and could claim one small record: she was the first vessel to shorten the passage from the Cape of Good Hope to Sydney by sailing through the Bass Strait.

Four months after sailing from Sydney, the *Book of Bearings* filling rapidly with bearings taken during a viciously complicated survey carried out in shoal waters, racing tides, sweltering heat, mosquitoes, mud, and mangrove swamps, and now beset with coral reefs, Flinders ordered his consort to return to Sydney. During those four months and the survey of over a thousand miles of coastline the *Lady Nelson* had strayed like a wayward puppy, had lost dagger boards and anchors. "The *Lady Nelson* sailed so ill," wrote an exasperated Flinders, "and had become so leewardly since the loss of the main, and part of the after keel, that she not only caused us to delay, but ran great risk of being lost; and instead of saving the crew of the *Investigator* in case of accident . . . it was too probable we might be called upon to render her assistance."

The coral reefs they were entangled with were those of the Great Barrier Reef (Flinders coined the name for this famous tourist attraction). Two days after the *Lady Nelson*'s topsail had vanished below the horizon the *Investigator* found an opening (now known as Flinders Passage) in the reef and sailed into the deep waters of the Coral Sea.[2]

A month later, after threading a way through the reef-strewn Torres Strait, which separates Australia from New Guinea, and three weeks into the Gulf of Carpentaria survey, Flinders had the *Investigator's* hull surveyed, as his command was leaking like a sieve, at times as much as 12 to 14 inches of water an hour. The report, as he suspected, was ominous. The *Investigator* was rotting away like a carious tooth. Flinders, sweating in the Gulf's humidity, wrote the dreadful report into his journal: "The ship having before made ten inches of water an hour in a common fresh breeze, we judge from that and what we have now seen, that a little labouring would employ two pumps; and that in a strong gale, with much sea running, the ship would hardly escape foundering . . . but if she remains in fine weather and happens no accident, she may run six months longer without much risk."

Three months later, with most of his crew suffering from violent diarrhea and himself lame from scorbutic ulcers on his feet, Flinders called a halt to the survey and headed for Kupang in Timor. The *Book of Bearings* had now added close to another two thousand bearings to its pages.

The fever-ridden Dutch settlement at Kupang provided much needed supplies and water, but the water proved treach-

[2]The *Lady Nelson's* curious career continued. After losing her anchors a substitute was made of waterlogged wood from an ironbark tree. But on the passage to Sydney this unique anchor dried out, the end result being surprise on the *Lady Nelson* and much hilarity among Sydney's spectators when the brig came to anchor. The sight of a floating anchor can be considered unusual.

erous. On the passage back to Sydney five men died from fever and dysentery; another four died in Sydney's hospital.

Flinders' *Book of Bearings*, although it helped produce his chart of the Gulf of Carpentaria (a chart not superseded until 1912) and solve one part of the riddle of compass deviation, could now claim a total of nineteen deaths in its making.

Chapter 13

THE FLINDERS BAR

Flinders' immediate concern after arriving at Sydney was to send his sick men to the hospital. Three days after they had been safely housed a report on his equally ailing ship confirmed that the *Investigator* was rotten and unseaworthy. Indeed, it was somewhat to the inspecting officers' collective surprise that she had managed to survive the passage from Timor to Sydney. But, much to Flinders' chagrin, the colony held no suitable vessel to continue the survey.

He also heard that the French had spent months in the colony, not sailing from Sydney until November, some four months after Flinders had sailed for the Gulf of Carpentaria.

With Flinders' survey falling apart due to the lack of a suitable vessel, he and Governor King concocted a plan. Flinders would sail for England as a passenger aboard the HMS *Por-*

poise; with him would go the precious draft charts, journals, and *Book of Bearings*. After arriving in England, Flinders would press the Admiralty for another surveying vessel to replace the *Investigator*.

On August 10, 1803, the *Porpoise* sailed for the harbor's entrance, her decks cluttered with the colony's senior officials. Just before entering the Pacific's swells the *Porpoise* hove-to, letters and dispatches for England were handed over, the crew gave three cheers, and the colony's officials scrambled down into boats to return to Sydney Cove and the routine of their lives.

Sailing with the *Porpoise* were two merchant ships, the *Cato* and the *Bridgewater*. A week later the *Porpoise* and the *Cato* lay mastless hulks being pounded to pieces on a coral reef some 750 nautical miles from Sydney. Three men drowned (one of them a boy who had been shipwrecked every time he went to sea, and called himself a Jonah), but ninety-four men struggled ashore to reach the comparative safety of a low-lying sandbank (later named Porpoise Cay) which lay behind the reef. The *Bridgewater*, having escaped the reef, made but little effort at rescue, and, thinking that all had been lost, blithely sailed on to Bombay.[1]

Flinders, as senior naval officer, organized a salvage operation on the *Porpoise*'s shattered hull. Included among the water and provisions were his draft charts, journals, and *Book of Bearings*, all soaked with seawater. He also decided to take the largest cutter, renamed the *Hope*, and sail for Sydney to organize a rescue effort. The men left on Porpoise Cay, with that infinite resource common to sailors, would make two small

[1]Her third mate and some other officers, appalled at their captain's behavior, quit the ship in Bombay. After sailing for England from Bombay the *Bridgewater* was never heard from again.

vessels from all the wreckage for their own rescue in case the *Hope* failed to arrive at Sydney.

Ten days after the shipwreck, Flinders, with Captain John Park of the *Cato* and twelve crewmen, sailed for Sydney. But before sailing, Flinders had the dried and salt-stained charts, journals, and *Book of Bearings*, packed away in casks and secured on the cay's highest point.

Twelve days later, two ragged figures disturbed Governor King at dinner with his family. King at first failed to recognize these strange apparitions in their salt-encrusted clothes, their bearded faces flayed red from sun and wind. They were Flinders and Park.

A few days later the two men were at sea again, bound for Porpoise Cay with a flotilla of three rescue craft. One of them was the 29-ton, locally built armed schooner the *Cumberland*. Launched in 1801, her original purpose had been to chase convicts attempting to escape in stolen boats. Her new purpose was now, with Flinders in command, to sail for Porpoise Cay, pick up the charts and journals, and then sail through the Torres Strait for England. It was, to say the least, an ambitious voyage for a small craft only 45 feet in length.

Flinders, a few days into the passage to the cay, soon came to realize the *Cumberland*'s disadvantages. Not only was she crank, steered poorly, and leaked so badly that she had to be pumped every hour, but she was also infested with rats, cockroaches, lice, and fleas. His method of writing his journal was to sit on a lee locker with his knees serving as a table, while bilge water sloshed across the cabin sole and seawater cascaded down from a broken skylight.

Six weeks after sailing from the cay, Flinders waded ashore to much cheering, handshaking, and an eleven-gun salute from carronades rescued off the *Porpoise*, and found that the castaways had built a rakish schooner and laid the keel of another. "It was," he wrote, "one of the happiest moments of my life."

Three days later, with all the saved documents stowed in a sea chest, the *Cumberland* weighed anchor and bore away before the prevailing southeasterly winds, bound for the Torres Strait and the first leg of the long passage home.

THE *CUMBERLAND*, UNKNOWN to Flinders, was sailing into a war zone. The Peace of Amiens had been scattered to the four winds and Britain and France were once more at war. And the war, under the bellicose Bonaparte (soon to be crowned Emperor Napoleon), promised to be brutal. A straw in the wind, showing the new direction of warfare, came with one of Bonaparte's first acts: the order to intern all British citizens found in France aged between eighteen and sixty.

After calling at Timor to water and a fruitless attempt to repair the overworked pumps and find pitch for the schooner's leaking deck seams, Flinders headed out into the Indian Ocean and the 6,000-mile passage to the Cape of Good Hope. Three weeks later, with the *Cumberland* leaking in a most alarming fashion, and with only one asthmatic pump working around the clock, Flinders made the fatal decision to alter course for Île de France and seek assistance from the French.

Into his journal went his primary reasoning for this decision, plus "several subordinate reasons of much less consequence." But one of them was to prove of great consequence. For Flinders had written, in all innocence, the following:

"Acquiring a knowledge of the periodical winds and weather there; of the port and the present state of the French colony; and how far it or its dependencies in Madagascar may be useful to Port Jackson; as also whether it may not be a convenient place for me to touch at during some part of my future expected voyage."

Flinders had no chart of the island and had to fall back on a gift from Sir Joseph Banks: a copy of the *Encyclopaedia Britannica*. From this Flinders gathered that the island's principal harbor, Port Louis, lay on its northwest side.

The island was raised at dawn on December 15, 1803. And three days later Flinders found that he had exchanged the discomforts of the wet and verminous *Cumberland* for the discomforts of a flea-ridden truckle bed in a humid room humming with mosquitoes, and a sentry pacing outside a locked door. He was a prisoner of the French.

Île de France, contrary to expectations, had turned into a trap. France and Britain, he learned to his cost, were again at war. And the island's governor, the formidable General Charles Decaen, a Bonapartist to the core, was a man who detested the English. Flinders' passport, Decaen noted, was made out for the *Investigator*, not the *Cumberland*. It was therefore invalid. Flinders' charts, journals, and log books were confiscated and Decaen's baleful eyes had soon lit upon that fateful December 4 entry. Flinders was obviously a spy. That the French had used their passport and been treated with great kindness at Sydney—and Decaen was well aware of this fact—was of no account. He was also aware that his countrymen had spent time spying—even making a chart with soundings of Port Jackson—and assessing the small colony's pathetic defenses. But this for Decaen, the Revolutionary lawyer turned soldier, was the realpolitik of the new century epitomized by his beloved Emperor Napolean.

Decaen, impervious to all pleas, was to hold Flinders a prisoner on the island for over six years. And, over the months and years as the Cumberlands were exchanged in prisoner cartels, Flinders became convinced that his nemesis lay in the short and stocky body of the island's malignant governor.

Flinders, during those years, his health poor, his eyesight failing in his right eye, fell into frequent bouts of depression. What sustained him was the drawing together of all the strands of his survey work in the *Investigator*. A number of those salt-stained books and charts were returned to him (having been inspected by the French) but, in the hot and humid climate of the island's coast, they soon turned into pulp and had to be transcribed and redrawn. Included among these soggy volumes was the invaluable *Book of Bearings*.

He also wrote. One paper, "Concerning the Differences in the magnetic Needle on board the *Investigator*, arising from an Alteration in the Direction of the Ship's Head," was sent to Banks and read before the Royal Society; and then published in the *Philosophical Transactions*. A year later another paper arrived on the barometer and its use in predicting weather changes at sea—a pioneer work in marine meteorology.

After two years, Flinders' captivity was eased by his giving Decaen his parole, and a promise not to walk more than 6 miles from new lodgings set in the middle of the island some 1,200 feet above the sweltering coast.

MAURITIUS (TO GIVE the island its present name) is a moodily beautiful island of volcanic origin, and one with a highly improbable history. Lying within the tropics and only some 40 miles across at its widest point, it has a coast rimmed with travel-brochure sandy beaches and coral reefs set in an aquamarine ocean. Uninhabited when first discovered, it was first settled by the Dutch in the seventeenth century and named Mauritius in honor of Prince Maurice of Nassau. The Dutch promptly set about killing for the pot what has become the world's most famous extinct bird: the plump, succulent, flightless, turkey-size *Didus ineptus,* also known as the Dodo (from the Portuguese word *duodo,* silly). African slaves bought from the Arabs worked on sugar plantations and logged ebony forests. Freed in this tropical heaven from the strictures of Lutheran pastors and the leaden skies of their homeland, the Dutch settlers slid into a delightful and slothful life of gargantuan eating and drinking. Governor after governor complained that the settlers had made gods of their bellies. The island was finally abandoned in 1710 for the harbor at Cape Town. And Mauritius turned into a safe haven for pirates waiting in ambush for the rich prizes in the shape of Dutch, English, and French merchant ships returning to Europe from India and the East Indies. Five years after the Dutch had abandoned the island the French laid their claim to it and changed its name to Île de France. Six years later they arrived with settlers.

The French fortified the island and used it as a springboard in their attempts to overthrow the English in India. The island's main revenue, in the years before Flinders' captivity, came from a group of famous French corsairs, mainly Bretons, led by Robert Surcouf, who pillaged the laden East Indiamen. French history has painted them in glowing colors as the romantic highwaymen of the Indian Ocean, but for British ship owners and insurers they were nothing but an expensive abomination.

FLINDERS' MOVE FROM the coast to the island's high plateau perhaps saved his sanity. It certainly improved his health. Le Refuge, the plantation home of the widow D'Arifat and her amiable family where Flinders was lodged, was set in the Wilhelm Plains with rivers running either side of the house and its outbuildings. Flinders found himself in a country of swift-running streams set in flower-covered banks, cascades, and waterfalls, improbable peaks rising into the mists, mimosa and mangoes lining winding tracks, pandanus palms gleaming in the rain. All this romantic lushness, so in keeping with the literature and music of the times (Flinders was an accomplished flute player), was salve and balm to Flinders' tortured musings on his captivity and the vile Decaen.

In these surroundings, his life at Le Refuge reads like a Jane Austen-in-the-tropics novel. His day began with a bath in one of the rivers, followed by breakfast with Madame D'Arifat and her family of three sons (aged between seven and twenty-seven years) and three charming daughters (aged between thirteen and twenty-one years). In this hothouse, highly charged tropical setting, the twenty-one-year-old Delphine soon developed romantic yearnings for the dark-eyed, brooding Englishman. After breakfast Flinders worked on his charts, journals, papers, and the draft version of the *Investigator's* voyage, and studied French. Dinner would be at 2 p.m., again with the family. During the afternoon he examined the English written by the two eldest daughters, and they, in turn, examined him on his French, followed by readings in both languages. Exercise came with walks in the surrounding countryside. The evening, after tea, was spent in playing picquet, whist, tric-trac (a form of backgammon), or music, or in conversation. Delphine adored playing tric-trac with Flinders. The day ended with supper, and all were in bed by ten o'clock.

It was in these congenial surroundings that Flinders continued his research into ships' magnetism. In his paper read before the Royal Society he had suggested that the focal point of attraction for the compass needle lay at the ship's center, where the cannon shot was stowed. But he was gradually coming to the conclusion that every item of iron was of influence—thus making the focal point different in each ship. Find this focal point and then the influence could be neutralized with a "counter attractor"—a vertical iron bar placed close to the binnacle.

BY THE SPRING of 1810, with a British combined-operations force closing in on Île de France, Flinders finally obtained his release from Decaen in an exchange of prisoners. Much speculation has surrounded this sudden release, the cynical claiming that Decaen, knowing he had little hope of success against the formidable force of ships and soldiers preparing to take the island, had come to a *quid pro quo* with the British. Flinders (whose long captivity had become something of a *cause célèbre* in Europe) would be released by Decaen: but in return the British would give generous treatment to the defeated French.

The British did attack. Decaen did surrender. And, much to the disgust of many British officers, the French soldiers and sailors were all repatriated to France; with Decaen and his staff being given passage home in a British transport vessel.

But all this was to happen months after Flinders sailed from the island as a free man. On that day in the June of 1810, all that Flinders could savor was sheer delight: "After a captivity of six years, five months and twenty-seven days, I at length had the inexpressible pleasure of being out of the reach of General De Caen."

FLINDERS' LONG ODYSSEY had come to a close. But back in England, together again with his wife, Ann, he found that although his paper on the magnetic needle had roused much interest in scientific circles, it had been met by the Admiralty with torpid indifference.

Much had to be done before he could tackle the problem of ship's magnetism, however; the comparative languor of those last few years at Le Refuge was over and he plunged headlong into a maelstrom of activity. Lodgings had to be found in London for himself and his wife; family friends had to be visited; the snake pit of finances, after years of absence, had to be resolved; countless meetings had to be held—with Aaron and John Arrowsmith, printers of charts and maps, on the matter of size and number of charts to be engraved; with Banks and the Admiralty on writing the narrative of his voyage; with Captain Thomas Hurd, Hydrographer of the Navy.

A few months after his return Flinders had the singular satisfaction of listening one morning to guns being fired to celebrate the fall of Île de France. Later the same day he was reading news of the victory in *The Times* and the *London Chronicle*.

In the early spring of 1812 a chance meeting with Ralph Walker—whom he had not seen for eighteen years, but whose azimuth compass he had used aboard the *Investigator*—led him to renew his investigations into compass deviation.

A few weeks after that meeting Flinders wrote to the Admiralty requesting that magnetic experiments be made aboard naval ships at Sheerness, Portsmouth, and Plymouth. It was a detailed letter written with the hope of testing his theories on compass deviation. The letter described the manner of swinging the ship to obtain a round of compass bearings on prominent marks ashore, with all these bearings to be taken from six different positions aboard the ship. The Admiralty, with surprising speed, agreed to all these proposals and ordered Flinders to supervise the observations.

By July, his report lay with the Admiralty. But it lay unread, for it presented the bureaucratic mind with a problem. To whose office should it be channeled? Such vacillating led an angry Flinders to write an acid, and uncharacteristic entry into his Journal: "No person at the Admiralty feels himself competent to form an opinion upon the subject; so that nobody will interfere, and the discovery runs the risk of being neglected."

The report, eventually, was read and digested. But the Admiralty thought it far too rich fare for His Majesty's naval officers, for it contained a bitter critique of the Admiralty and its dockyards when it came to the selection and care of compasses. All these instruments were bought by contract and never inspected. To Flinders, this was culpable folly, and he suggested that an Inspector of Compasses be appointed. Matters were just as bad aboard ships: "The Boatswain of the ship, who has nothing to do with the accuracy of the course, receives the compasses into his charge; and puts them away with other things, into his store or sail room; or if he is a careful man and has the means, he places them upon a shelf or in a locker in his cabin; perhaps with his knives and forks and a few particular marline spikes."

Flinders, in his report, arrived at the conclusion that he had been mulling over since Île de France: Compass deviation was dependent on the direction of the ship's head and was also proportional to the magnetic dip. The ship's magnetic latitude, in other words, also affected compass deviation. To counter the deviation caused by the ship's iron work he suggested that "a strong bar of old iron, of such length as that when one end is let into the deck, the other will be nearly upon the level with the compass card."

The Admiralty requested Flinders to write an abridged version of his long report. This was then printed and distributed to all captains and commanders. It also appeared in the *Naval Chronicle* of 1812 under the title of "Magnetism of Ships."

Flinders never held, or saw, the final printed version of his narrative story of the *Investigator*'s voyage. The last year of his life was spent in acute pain with a bladder and kidney inflammation, his last days in a coma. Ann Flinders thought her husband, in the last year of his life, though only forty, looked like an old man of seventy.

A Voyage to Terra Australis was published on July 18, 1814, with a copy being delivered to Flinders' London lodgings. He was in a coma, and the two volumes, fresh and crisp from the printers, were placed on his bedside table. He died early next morning.

In a lengthy appendix to his *Voyage* he again suggested the use of a vertical iron bar to correct compass error. Thus, what had been born as an idea on the heights of Mauritius was eventually to become an essential component attached to the binnacles of thousands of ships. But it was to take many years and many shipwrecks before his idea, the Flinders bar, came into use.

Chapter 14

SOFT IRON, HARD IRON

Sir Joseph Banks, overweight, gout-ridden, and confined to a wheelchair, outlived Flinders by six years. But Flinders was not the only seaman to whom Banks had extended a helping hand and befriended with his patronage. Another such man was William Scoresby, who was whaling in the Arctic when Banks died in 1820. That the Royal Society's president should have invited a whaler to his celebrated breakfasts and dinners, maintained a twelve-year correspondence with him, and then introduced him to influential men of science, is a cause for wonder. But Banks had a keen nose for the unusual intellect. And Scoresby was a most unusual man.

The son of a very successful Whitby whaling captain (also a William and, like the Deity, always capitalized as Father by his son), Scoresby had been baptized into the rigors of Arctic whaling (Father being the captain) at the age of ten. Three years after this precocious and frigid initiation, Scoresby's education had settled down into a pattern. Summers were spent whaling in the Arctic, winters in more formal education at school in Whitby and then, in 1806, at Edinburgh University.

The summer prior to entering university he had sailed as chief mate aboard his father's whale ship, the *Resolution*. On this voyage, sailing through loose pack ice north of Spitzbergen, they had reached 81°30' North, a unique situation that made the *Resolution*'s crew the most northerly people in the world. In these high latitudes the Scoresbys noted the sluggish behavior of the compasses and the large magnetic variations. It was this, no doubt, that caused Scoresby to write to his son at Edinburgh that he should take every "opportunity to know the best method of giving power to the magnet and charging the needle of the compass." But Father was in the habit of bombarding his son with parental advice. The ten volumes of Shakespeare sent to him at university had not been provided for William to gain "a thirst for the theatre," but to learn from the "language that runs through the whole," even though the language was "poor in itself compared with divine writings."

At Edinburgh the dark-eyed Scoresby, tall, slight of build, with curling dark hair framing a mobile and intelligent face, came under the wing of Professor Robert Jameson. The professor, intrigued by having a whaler as a student, encouraged Scoresby in his study of the Arctic's natural history. One of the unforeseen results was that Jameson, much to his perplexity as to the housing and feeding of the animal, received a polar bear as a present. The other result was the publication, in 1820, of Scoresby's *An Account of the Arctic Regions*. The two volumes, which are considered the foundation stone of Arctic science, were dedicated to Jameson.

All aspects of the Arctic, from plants to the sea, came under the scrutiny of Scoresby's inquiring mind. He was the first to describe and make accurate drawings of snow crystals (unsurpassed until the invention of the photographic microscope: ninety-six crystals, all different, are illustrated in the *Arctic Regions*). He fashioned lenses from ice, using first an ax, then a knife, followed by a polishing with his hands. With the lens he fired gunpowder, melted lead, burned wood, and, much to their astonishment, lighted his crew's tobacco pipes. He invented an instrument which he called a "marine diver" to take the sea's temperature at various depths. He collected and sketched the minute plankton, visible only under a microscope, of the polar sea. Such life he knew was essential to the polar region: "And thus we find a dependent chain of existence, one of the smallest links of which being destroyed, the whole must necessarily perish." He noted that eating polar bear liver was dangerous; for sailors who ate the liver either died or suffered from having their skin peel away in shreds.[1]

On one voyage, his ship frozen into thin ice through which whales were surfacing to breathe, he devised a most unique method of whaling: "I provided myself with a pair of *ice-shoes*, consisting of two pieces of thin deal, six feet in length, and seven inches in breadth. They were made very thin at both ends; and, in the center of each, was a hollow place exactly adapted for the reception of the sole of my boot, with a loop of leather for confining the toes." Sliding these primi-

[1]Polar bear liver contains massive, fatal amounts of vitamin A. It was eating polar bear liver that killed all the members of a Swedish Arctic expedition in 1897.

tive skis (Scoresby happened to be totally unaware of this Norwegian invention) across the dangerously thin ice, he harpooned and lanced three whales.

Scoresby also dealt with the problems of the magnetic compass in a seventeen-page appendix to the *Arctic Regions*. Entitled "On the Anomaly in the Variation of the Magnetic Needle, as Observed on Ship-Board," the appendix was a paper that had been sent to Sir Joseph in 1818, and read before the Royal Society on February 4, 1819.

Scoresby's investigations into compass deviation—the anomaly of the paper's title—had been based on measurements taken aboard the whale ship *Esk* during the summer of 1817, a summer in which the *Esk* had sailed far north along Greenland's east coast in an open sea—a sea usually inaccessible due to pack ice. Scoresby, on his return to England, made the disappearance of the ice known in the newspapers. This news brought a letter from Sir Joseph asking for more information. Scoresby's reply was prompt:

> I found on my last voyage about 2000 square leagues of the surface of the Greenland Sea, included between the parallels of 74° and 80° north, perfectly void of ice, which is usually covered with it. Now all this ice has disappeared within the last two years, and there is little doubt but it has been drifted to the southward into warmer climes, and there dissolved. . . . Had I been so fortunate as to have had the command of an expedition for discovery, instead of fishing, I have little doubt but that the mystery attached to the existence of a northwest passage might have been resolved. There could have been no great difficulty in

exploring the eastern coast of Greenland, and probably the fate of the colony established by the Icelanders so many centuries ago might have been ascertained. I do conceive there is sufficient interest attached to these remote regions to induce Government to fit out an expedition, were it properly represented. . . . . I should have much satisfaction in attempting an enterprise of this kind, namely to examine the islands of East Greenland or Spitzbergen, especially the eastern part, which has not been visited for many years past.

Further letters followed between the two men, with Scoresby listing the geographic, commercial, and scientific objectives of such a voyage. The science included terrestrial magnetism with observations of magnetic variation, dip, and intensity, along with investigations into compass deviation.

These letters led—for they were passed to the Admiralty by Banks—to the renewal of the English obsession with the search for a northwest passage from the Atlantic to the Pacific. It also led to a curious meeting at Banks' Soho Square house between Scoresby and John Barrow, the Second Secretary of the Admiralty. Barrow, a powerful and devious bureaucrat, had taken up Scoresby's ideas for a polar expedition (claiming it as his own) and had hinted to Banks at some position for Scoresby. Barrow, on meeting Scoresby, proved evasive and rude. The Arctic expedition, it appeared, would be led by Royal Navy officers with zero knowledge of Arctic ice conditions. And an embarrassed Banks admitted to Scoresby that all the officers had been appointed, but that a position "in some subordinate capacity" could perhaps be arranged. Scoresby, from first-hand knowledge of the Royal Navy, refused.[2] What also rankled with Scoresby, a frugal Yorkshireman, was that he had to pay out of his own pocket for the return travel between Whitby and London.

SCORESBY'S INVESTIGATIONS INTO compass deviation came very close to revealing the true law which Flinders had missed. Scoresby noted that a large quantity of iron in a wooden ship lay in a vertical position: hanging knees, nails, and bolts in the deck, the capstan spindle, anchor flukes, stanchions, chain plates, belaying pins, rudder stock. All these vertical iron items became magnetized by the earth's magnetic field—an effect now known as induced magnetism or magnetic induction. In the northern regions the upper ends of the vertical iron showed *south* polarity. He demonstrated this effect aboard the *Esk* by placing a compass upon the capstan head, just off-center from the iron spindle, so that the north point of the compass was attracted by the spindle's upper, south polarity, end. And by moving the compass around on the capstan he astonished his officers by making the ship apparently to be "steering a course directly contrary to that intended."

He also deduced, and this added to Flinders' findings, that the increase in deviation observed aboard ships in high latitudes was due to the combination of two magnetic forces: the vertical and the horizontal. As a ship sails into the high latitudes she is also sailing into a region of increasing dip. At the

---

[2]In 1807, Scoresby had volunteered, along with other Whitby whalers, to help sail back to England Danish vessels captured at the Battle of Copenhagen. His experience of the floggings, drunkenness, and poor seamanship in the Navy made a permanent impression on him.

magnetic north pole a dipping needle stands vertical. This vertical magnetic force increases the magnetic power of the ship's vertical iron. But, as the vertical component of the earth's magnetic field increases, the horizontal component decreases, thus lessening the directive force of the magnetic pole on the compass needle.

Like Flinders, Scoresby found that the greatest compass deviation came with the ship steering either east or west, the least when steering north or south.

Large deviations, as Scoresby pointed out, could lead to unpleasant surprises. If a ship with a 10-degree deviation (and the navigator not allowing for this) were to steer west for 300 miles, turn about, and then steer 300 miles east, she would not find herself back at her starting point, as one would suppose, but 104 miles to the southward.

Scoresby's selection of shipboard iron is known as "soft" iron—items easily magnetized by the earth's magnetic field but which, once the magnetic field is removed, lose their magnetism. Thus, all those vertical soft iron items aboard a ship sailing from the northern hemisphere into the southern hemisphere will gradually lose their magnetism as the ship approaches the magnetic equator—where the magnetic dip is zero—and then gradually regain it. But their polarity will be reversed: in the southern hemisphere the upper ends will have changed to a north polarity. The north point of Scoresby's compass on the capstan head would now be repelled by the capstan's spindle, not attracted.

KNOWING THE REASON for the antic behavior of a compass needle is one thing. For the careful navigator to make cor-

rect adjustments in his calculations is another. Most navigators by Scoresby's day had come to terms with compass variation, its measurement, and the correction to be applied in
their workings. But the Admiralty had neglected Flinders'
suggestion of placing a vertical iron bar close to the binnacle
to compensate for one aspect of compass deviation: nor, for
that matter, had any merchant-ship owner or captain followed his suggestion.

However, at least for the scientific navigator, help was at
hand. In 1820, the same year as the publication of Scoresby's
Arctic Regions, a paper was published in *Brande's Quarterly Journal* on aspects of compass deviation by a certain Dr. Thomas
Young. Young was a most remarkable man, with a bulging
quiver of scholastic arrows, and known at Cambridge University as "Phenomenon Young" for his vast erudition. He has
been called the "founder of physiological optics," he coined
the word *energy* for the power of doing work possessed by a
moving body by virtue of its motion (kinetic energy), makes
a cameo appearance in every textbook on the strength of
materials with what is known as Young's modulus, discovered
the cause of astigmatism, and established the wave theory of
light. He was a lapsed Quaker who enjoyed music and dancing, his linguistic skills included Latin, Greek, Hebrew,
Chaldee, Syriac, Persian, German, French, Italian, and Spanish, and he had made a fundamental contribution to the deci-
phering of the inscriptions on the Rosetta Stone. He also
happened to be Superintendent of the *Nautical Almanac*, and
Secretary of the Board of Longitude, and had written a
report for the Admiralty on shipbuilding. He was, in other
words, well known to the Admiralty, and he sent them his
paper on compass deviation.

THE FOUR ROYAL NAVY ships sent into the Arctic for the summer of 1818 had returned. The two-pronged attack—two ships attempting to sail to the North Pole and two ships attempting to discover a northwest passage—had failed. And all the ships had been bedeviled by the compass needles' erratic behavior. On two ships the compasses differed by as much as 11 degrees, so that sailing parallel, as indicated by compass signal, was impossible. On another ship a bearing taken on a point of land before tacking was southeast, but after tacking the bearing suddenly changed to south. Captain John Ross, commanding the northwest passage probe, coined the word *deviation* to describe what had before been known as local attraction or magnetic anomaly. Ross thought, incorrectly, that deviation was affected by cold, heat, and wind.

The idea behind Young's paper, using the magnetic data gathered by the Arctic ships, was to see if the errors of a ship's compass—the deviation—measured at one place, say London, could be forecast when the ship sailed north into polar latitudes. Young's calculations produced an answer, hidden in a thicket of formulae: "Table of Corrections for clearing the Compass of the regular effect of a Ship's Permanent Attraction."

Before sailing from England the ships of the Arctic expedition had all taken measurements of compass error and, using a dipping needle, magnetic dip. During their voyage into the Arctic these measurements had continued. Young's theory and calculations, contained in his Table of Corrections, agreed with the observations made aboard the ships. The only drawback to this most satisfying result was that the magnetic dip had to be measured before the deviation could be obtained from the tables. And dipping needles—except on scientific expeditions—were not part of standard navigating equipment.

What Young had done was to grasp the concept of dividing a wooden ship's magnetic material, for the purpose of compass correction, into "soft" iron, which soon lost its magnetism, and "hard" iron, which retained its magnetism, both of which had to be allowed for. With this paper Young laid down the basis for the modern theory of a ship's magnetism and its compass correction. But, as so often with the Admiralty, the paper vanished into a bureaucratic black hole—and it is doubtful if any shipowner was aware of Young and his theories.

Flinders' suggestion of an iron bar had vanished into this same black hole when Professor Peter Barlow, of the Royal Military Academy, produced a method of correction using a vertical iron disk attached to the binnacle. The disk, known as Barlow's plate, was fiendishly difficult to install in its correct position, sometimes obscured the compass, and with its dangerous spike sprouting from its center, made life hazardous for crew and helmsman. A number of vessels were fitted with this strange contrivance. But a greater hazard than its wicked spike soon emerged: the Barlow plate might correct for the compass deviation at the port where it was installed, but on vessels sailing into different magnetic latitudes—particularly those sailing across the magnetic equator—the compass error, much to the navigator's dismay, was found to increase. Such news, traveling the waterfront, effectively killed Barlow's plate.

Barlow also made a report on the state of the Admiralty's compasses in store at Woolwich. "More than half of them would be considered *mere lumber*," wrote an appalled Barlow, "which ought to be destroyed on the same principle as we clip base coin, being wholly useless while in store and *extremely*

dangerous if suffered to pass out of it." Such compasses, continued Barlow, were "wretchedly defective" and would have "dis-
graced the arts as they stood at the beginning of the 18th
century."

This report also fell into the vast Admiralty black hole. Two
years later, again at the request of the Admiralty, Barlow made
another inspection to find that nothing had changed. "It is
constantly to be regretted," sighed an amazed Barlow, "that
while all the other appointments in the British Navy are
so excellent, the compass (an instrument of unquestionable
importance and the least expensive in ship), should be so very
inferior as it is at present."

THE SAME YEAR as Barlow's second report, 1822, a most curious looking vessel sailed from the Thames and across the
Dover Straits to the mouth of the River Seine, and then up
the river to Paris. The *Aaron Manby*, her 3-foot diameter, 47-
foot-high funnel pouring out smoke, cinders, and smuts,
paddle-wheels thrashing the murky waters of the Thames into
a creamy froth, was making history by being the first iron-
hulled vessel to cross a sea. The iron sheets of her hull plating
and frames had been made in Staffordshire, and then riveted
together in a dock on the River Thames. For her first voyage
she carried a cargo of linseed oil and iron.

She also carried within her, unseen and unknown, a type
of magnetism common to all iron-hulled ships: the permanent
magnetism of "hard" iron—iron that absorbs magnetism by
being pounded and riveted, and retains its magnetism. An iron
ship, in other words, is a permanent, giant magnet. But, just to
confuse the compass even further, the *Aaron Manby* still had

her soft iron magnetism in her rudder stock, stanchions, and most spectacularly in that prodigiously high funnel.

This combination was to prove a menacing and dangerous combination. In the decades after that first voyage of the *Aaron Manby*, more iron-hulled vessels were launched from builders' yards, and a pattern began to emerge that alarmed sailors, ship owners, insurance underwriters, and shippers, who refused to have their cargoes put aboard iron-hulled ships. For these ships, it appeared, had a most distressing habit of being led astray by their compasses. If they were lucky, they survived. If unlucky, they ended up being stranded or wrecked.

Flinders, Young, and Scoresby had based their magnetic theories on sailing ships with wooden hulls. Such vessels have negligible amounts of "hard" iron compared to their "soft" iron. The building of ships with iron hulls and steam engines changed this magnetic proportion. And the resolving of these vexing magnetic proportions, over the next five decades, was to cause much controversy and argument.

Chapter 15

"An Evil So Pregnant with Mischief"

In the chill dawn of a June day in 1831 the *Lord Dundas*, an iron-built steam vessel, churned down the River Mersey from Liverpool and out into the Irish Sea. Just a few days previous to this departure, during a trial run from Runcorn to Liverpool, with the *Lord Dundas* throwing spray and thumping through the river's short chop, the engineer, George Rennie, alarmed at her apparent fragility and hearing that she was about to sail for the Clyde, advised all who sailed with her to wear cork life jackets.

Such a view, perhaps, was not surprising. The *Lord Dundas* had been built in Manchester to steam along the canal between Glasgow and Edinburgh. Her iron hull was 68 feet long, her power a steam locomotive engine which turned a 9-foot-diameter paddle wheel set along the centerline in a trough. She only weighed 7.8 tons and could happily float in 18 inches of water.

Her builder and designer, William Fairbairn, perhaps heeding Rennie's advice, was not aboard his creation when she left the Mersey. He followed her that afternoon aboard the steam packet that ran between Liverpool and the Isle of Man. Here, Fairbairn would meet and inspect his vessel after her 70-mile passage. But on arriving in the Isle of Man he found no *Lord Dundas*. Nor, for that matter, had anyone seen or heard of her. And so an agitated Fairbairn made a pier-head jump aboard a steamer heading for the Clyde. At Greenock, heads were shaken when he asked if the *Lord Dundas* had been seen. Considerably worried, for he now thought himself responsible for six deaths, he hired a small boat and searched among the Cumbrae Islands in the Firth of Clyde. Again, nothing had been seen and no wrecks reported.

As a final resort he returned to Douglas on the Isle of Man. Here he heard that a vessel answering his description was anchored off Ramsey, some 12 miles away. Hiring a horse he rode for Ramsey and, as he breasted the last hill, he spotted the *Lord Dundas* riding at anchor on a sea as smooth as glass. But his chase was not yet over. A sailor lounging on the waterfront told him that all the crew were drinking in a local inn; but at the inn he was told that all the crew had moved on to a country fair some miles inland. An exhausted and angry Fairbairn sat down to wait for their return and an explanation for their mysterious vanishing and reappearance.

On the crew's mellow return from the fair, and having heard their tale, it became obvious to Fairbairn that the *Lord Dundas* had been led astray by her compass. For the compass course steered from leaving the Mersey, instead of bringing them to the Isle of Man, had brought them to the Cumberland coast (a frightening error of over 50 degrees), where they had spent the night anchored in Morecombe Bay. In other words, the compass might have shown them to be steering NW by W, but they were, in fact, steering North. From Morecombe Bay, hazarding a guess at the course, they had steamed to the Isle of Man.

Fairbairn, before sailing for the Clyde in the *Lord Dundas*, made an effort to correct this huge deviation caused by the iron hull and engine. Another compass was placed on shore, and then a piece of iron was moved around the canal boat's compass until both compasses agreed on direction. "With this rough-and-ready correction," wrote Fairbairn, "we proceeded on our voyage with perfect certainty and without further mishap."

THE CREW OF the *Lord Dundas* had been lucky in their erratic course across the Irish Sea. Other men and women were not so lucky when it came to compasses being led astray by iron. For in the early decades of the nineteenth century wooden vessels were having more iron incorporated into their fabric and fittings: internal diagonal bracing, fastenings, knees, stanchions, davits, water tanks.

In the March of 1803 the newly built thirty-six-gun frigate HMS *Apollo* sailed from Ireland escorting a convoy of seventy merchant ships. A week after sailing, in the dark hours of the early morning, the frigate and forty of her convoy piled ashore on the coast of Portugal just north of Cabo Mondego. The other thirty of the convoy, disagreeing with the *Apollo*'s evening-signaled compass course, had sailed a different course and so escaped shipwreck. The *Apollo*'s compass had fooled the unfortunate master into believing that they were sailing a safe course some 180 miles off the Portuguese coast.

Another Royal Navy vessel, HMS *Thetis*, sailed from Rio de Janeiro in 1830 carrying a million dollars in silver and other treasures. Within a day of sailing, thinking the ship well clear of Cabo Frio, the *Thetis* altered course in the night and ran at 9 knots into the vertical cliffs of an island close to the cape. Jib-boom and bowsprit snapped like carrot sticks, all the masts crashed overboard, the ship sank stern-first, and twenty-five men drowned.

A few years later the *Reliance*, an East Indiaman homeward bound up Channel with the master convinced they were close to the English coast, drove ashore on the French coast with a loss of 109 lives. She carried a 46-foot-long iron water tank which happened to be sited close to the binnacle.

All these disasters can be blamed on compass deviation. But the compass deviation aboard wooden vessels was minor compared to the startling deviations measured aboard the increasing number of iron-hulled vessels. One pioneer iron-ship builder was John Laird. Based at Birkenhead, on the opposite side of the River Mersey from Liverpool, Laird's first foray into this new material had been a 60-ton lighter for use on the Irish Lakes. This was followed by a paddle-wheel steamer for the Irish inland waterways; then the *John Randolph*, the first iron-hulled paddle wheeler to be seen in the United States. All these had been prefabricated and the parts shipped out to be assembled. But in 1832, a decade after the *Aaron Manby* had made her historic voyage, an iron-hulled paddle wheeler sailed from Liverpool bound for the west coast of Africa and the River Niger.

The voyage of the *Alburkah* was something of a family venture for John Laird. She had been designed and built by his younger brother for a trading venture into the interior of Africa, financed by a group of Liverpool merchants. The *Alburkah* was small, only 55 tons, with an iron hull and wooden deck. During the voyage out to Nigeria the surgeon who sailed with the expedition noted that the "compasses were useless from the attraction of the vessel."[1]

Such news, as this was the first oceanic voyage of an iron-hulled ship, was profoundly disturbing to Laird. He was banking his all on this new ship-building technology. And iron ships weaving across oceans with useless compasses were hardly the best advertisement.

But Laird was a shrewd and astute business man. In the building of an iron-hulled paddle wheeler for use on the lower reaches of the River Shannon and the west coast of Ireland, he saw his opportunity to enlist the support of a St. George to slay the dragon of compass deviation. And this St. George, urged into combat, flew a version of the saint's banner with its red cross on a white background: the White Ensign of the Royal Navy.

DURING THE AUTUMN of 1835 the iron-hulled, 210-ton *Garryowen*, with her two 85 horsepower engines turning paddle wheels 15 feet 6 inches in diameter, and with a 28-foot-high funnel sprouting between the two masts of her schooner rig, lay at Limerick on the River Shannon. Built by John Laird

---

[1] The surgeon was one of the luckier members of the expedition. Thirty-nine of the men who sailed from England died of malaria, only nine returning alive.

and owned by the City of Dublin Steam Packet Company, she sat waiting in the soft Irish rain for her side-whiskered savior in the shape of a half-pay commander of the Royal Navy. John Laird and the *Garryowen's* owners had managed to prod the Admiralty into action and forced Their Lordships to grapple with the alarming problem of compass deviation in iron ships.

The Admiralty picked Commander Edward Johnson, an officer with surveying experience and one who had also made experiments with compass needles, to head the investigation. Burdened with azimuth compasses, binnacle compasses, theodolites, instruments for measuring magnetic dip and intensity, cumbersome Barlow plates, and with his head full of sage advice from Professor Peter Barlow and Samuel Hunter Christie (a professor of mathematics at the Royal Military Academy), Johnson set off for Ireland and the waiting *Garryowen.*

With no wet dock available in Limerick for swinging ship, Johnson had the *Garryowen* moved to Tarbert Bay, where she was securely moored, and began his preparations for the measurements. Strange wooden structures appeared on the steamer: a fake poop to imitate those found on His Majesty's ships; a timber structure projecting from the stern over the water; platforms high on the foremast and on the foredeck. All these were to provide stable platforms for his array of compasses from which bearings would be taken on a nearby mountain peak. Over the next month, often having his experiments interrupted by gales and rain, Johnson found deviations of over 30 degrees, and found the Barlow plate incapable of correcting such massive errors.

But one of his most important findings was in an experi-

ment when the *Garryowen* was carefully warped toward compasses set up on a quay. As the ship was brought alongside he
found that the needles of the compasses, opposite the bow and
stern, were deflected in opposing directions. In other words,
the *Garryowen* was a 130-foot-long magnet with a north pole
and a south pole. This was not soft-iron magnetism, but hard-
iron magnetism: "As in the construction of iron vessels, hammering rivets, might elicit magnetic influences, it would be
well to note, by compass, the direction of their heads and
sterns when building, with a view of ascertaining whether . . .
any distinct magnetic properties indicated by those parts are
due to the line of direction of the vessel with respect to the
magnetic meridian." In short, the direction in which the vessel
lay during building could affect her magnetism. Johnson's
conclusion, most worrying to Laird, was that a compass position aboard a wooden vessel could not be duplicated aboard
an iron vessel. His only advice was to raise the compass above
the iron's influence.[2] His other suggestion, a most important
one and central to all later mechanical means of compass correction for deviation, was the thought that "it might be useful
to ascertain how far another magnet in a given position was
capable of correcting for the deflection."

Johnson made his report to the Admiralty in January 1836.
Two months later it became public knowledge when it was
read to the Royal Society by Captain Francis Beaufort, the
Admiralty Hydrographer, followed by its publication in the
Philosophical Transactions.

[2]Scoresby had made the same suggestion in his *Arctic Regions*: "Hence I
have been in the habit of carrying a compass, occasionally in the crow's
nest, fixed at the mast head, where it was found to be free from those
anomalies which are so sensible in a compass on deck."

Johnson, much to Laird's dismay, had failed to kill the dragon threatening his business. But help was at hand from Francis Beaufort.

IN THE PANTHEON of the many gods worshiped by the British Royal Navy, three figures stand out as paragons: James Cook as the exploring sailor; Horatio Nelson as the fighting sailor; and Francis Beaufort as the head of a hydrographic department.

Cook's London statue, with the Admiralty Arch on its right, gazes across the tourists hurrying down the Mall to Buckingham Palace. Nelson's, high on his column, presides over the tat and pigeons of Trafalgar Square. Beaufort has no statue, only his international wind scale: a method of describing wind and sea familiar to all sailors and listeners to shipping forecasts—a Force 10 for an English speaker being the same as a *dix Beaufort* for a French speaker.[3]

Beaufort was aged fifty-five when appointed as Hydrographer (the age at which today's Hydrographer is obliged to retire), and had been in this office for seven years when he read Johnson's report to the Royal Society. Those were the days when the Admiralty formed the principal conduit for the flow of government money into the sciences. The funds of the learned societies, the Royal Society, the Royal Institution, and the British Association were far too meagre to mount any major research. Science at Oxford was moribund and the universities of Cambridge, Dublin, and Edinburgh had only enough resources for home-based research.

[3]Force 10 on the Beaufort Scale has a mean wind speed of 54 knots and is described as a Storm.

Beaufort, with a youthful zest for the latest in technology, acted as the ideal middleman between the sciences and government funding. He was helped by the fact that he wore other hats than that of Hydrographer, for he was a member and officer of three important institutions—the Royal Society, the Royal Astronomical Society, and the Royal Geographical Society. Plus, it was his office that administered and controlled the finances of the Royal Observatory at Greenwich and its branch at the Cape of Good Hope.

A steam paddle wheeler had been used during the 1827–1828 survey of the River Thames, and the benefits of a steamship in survey work were obvious to Beaufort. Sounding lines could be run in a systematic way and surveying could continue in calm and windy weather. But the problem with the new breed of iron-hulled steamships lay in their compasses.

A year after reading Johnson's report to the Royal Society, Beaufort pointed out to the Admiralty the "many instances of HM Ships having been endangered and their services delayed through the badness of their Steering Compasses." It was a state of affairs, continued Beaufort, that had become "so notorious that it is a matter of surprise that more serious mischief has not been the result." He proposed a committee—giving a list of suitable names—to look at the problem and, as bait to the frugal Admiralty, suggested that the committee could operate without any expense to the country.

Within three days the suitable names received invitations from the Admiralty asking them to meet at noon on July 24, 1837, in the splendors of the Admiralty Library. There they would consider the defective state of the magnetic compass and find some remedy for "an evil so pregnant with mischief."

Only nine days had elapsed between the writing of Beaufort's minute to Their Lordships and the first meeting of what became known as the Admiralty Compass Committee, which suggests either a remarkable efficiency or, as is more likely, quiet spadework by Beaufort, a man with considerable charm, in sounding out his list of committee members at quiet dinners and in conversations at society meetings. One of the men happened to be Johnson.

THE YEAR OF the Admiralty Compass Committee's formation was also the year of the launching of the largest iron-hulled steamship so far ever built. Ordered by the General Steam Navigation Company, she was designed to run between London and Antwerp. The *Rainbow* was another John Laird creation, and with her speed—it was claimed she could steam at 14 knots—and 581 tons of elegant looks combined in her graceful sheer, bow and counter, tall funnel, and schooner rig, she was soon the talk of the waterfront. But on her maiden voyage from the Mersey to the Thames she had come close to shipwreck. Blanketed in thick fog off the Isle of Wight the *Rainbow*'s captain had set what he thought was a safe compass course clear of all dangers, but was warned by a passing fisherman that he was heading straight for the shore. Plans were soon in the making to deal with the *Rainbow*'s compass.

During the summer of 1838 Beaufort wrote to the Astronomer Royal, George Airy, at Greenwich Observatory asking him to inspect the compasses and magnetism of the *Rainbow*, even though the two men were not on the best of terms. The thirty-seven-year-old Airy had been appointed

Astronomer Royal in 1835 and had found, much to his surprise, that the Greenwich Observatory appeared to be a mere testing place for the Royal Navy's chronometers rather than a place of science. On finding a book of letter references he discovered that out of the 840 letters, 820 were related to chronometers. Darkly vowing to change this proportion, Airy had "mentally sketched my regulations for my own share in chronometer business."

For the rather austere and oracular Airy, regulations and order were the hinge upon which his whole life swung. Every scrap of paper had to be saved—check stubs, notes to tradesmen, circulars, bills, letters. Perhaps the saddest comment on this obsession came from his son: "As his powers failed with age the ruling passion for order assumed a greater prominence; and in his last days he seemed to be more anxious to put letters which he received into their proper place for reference than even to master their contents."

As a small boy at school in Colchester he had been noted for his skill in making peashooters and for his extraordinary memory where, at one examination, he recited 2,394 lines of Latin verse, the latter talent, perhaps, being of more use at Cambridge where he graduated as a brilliant mathematician. This was followed by his election as Lucasian professor of mathematics, Plumian professor of astronomy, and director of the Cambridge Observatory.

In 1838, at the urging of Airy, a Magnetic Observatory had been added to the buildings at Greenwich. So Beaufort's letter asking an Astronomer Royal to inspect a ship's compass is not as curious as it appears.

Airy had read Johnson's report on the *Garryowen* and was

aware of the problems afflicting the compasses of iron ships. He had also noted Johnson's tentative suggestion that the evil so pregnant with mischief, the dragon of deviation, could be slain with a magnet. Yet another Astronomer Royal, following in the footsteps of Halley, was about to make his contribution to nautical science.

Chapter 16

DEVIATION,
THE HYDRA-HEADED
MONSTER

Deptford and Greenwich sit cheek-by-jowl along the south bank of the River Thames. They also share a long maritime history. But Deptford's burden of history, compared to that of Greenwich with its background of royal palaces, parks, and the cool, graceful cluster of buildings by Inigo Jones, Sir Christopher Wren, Sir John Vanbrugh, and Nicholas Hawksmoor, is far less regal.

When Airy walked into the Royal Observatory at Greenwich as the seventh Astronomer Royal, Deptford had been dirtying its hands for over three hundred years as a dockyard, shipbuilding yard, and victualing yard for the Royal Navy. The sailor's sardonic epithet for the victualing yard, Old Weevil, gives a hint as to the content of their ship's biscuit and bread. But Deptford did have, for Airy, an ideal basin in which the *Rainbow* could be swung.

The steamer was brought into the basin and Airy walked around her and selected four positions for her compasses. The *Rainbow*, under the interested gaze of the dock workers, was then carefully swung to measure the deviations. They were massive: on some headings as much as 50 degrees. After days of work Airy had collected a mass of data, and then started working on the numbers to reduce them to some sort of order. He was now in his element. For Airy, the supreme methodical man, often spoke of mathematics as nothing more than a system of order carried to a considerable extent. But at first, much to his dismay, the numbers refused to assemble themselves into satisfying ranks. More numbers were required, this time by measuring the intensity of the horizontal magnetic field at each compass position and for each measured deviation. With these extra numbers, the ranks began to march, like obedient soldiers, to Airy's drum.

Within days, Airy was measuring, in his new Magnetic Observatory, the intensity of the magnets he was going to use for the compass correction. He was now ready for his grand and innovative experiment. On August 20, with workmen carrying magnets and scrolls of soft iron, he set about correcting the steamer's compasses according to his now compliant numbers. A 2-foot-long bar magnet and a scroll of iron were positioned to correct the main compass, and 14-inch bar magnets at the three other compasses. The *Rainbow* was swung, Airy peering anxiously through his spectacles as she made her slow and controlled circuit of 360 degrees. All was well. The steamer's compass, as Airy said with considerable pride, "was now sensibly correct."

A test sail on the Thames only reinforced this opinion. And then, courtesy of the grateful General Steam Navigation

Company, Airy and a party of friends made a celebratory cruise to Sheerness and back. On September 9 the *Rainbow*, with much fanfare, sailed on her maiden voyage to Antwerp with Airy aboard her for the first few miles down river. News that Airy, a slight, stoop-shouldered, and astigmatic St. George, had apparently slain the dragon of deviation soon swept through the shipping industry.

A month after the *Rainbow* had sailed on her first voyage to Antwerp, Airy received a letter from Nathan Cairns, a Liverpool shipowner who was about to take delivery of a new ship. He had heard of Airy's success with the *Rainbow*, wrote Cairns; would Airy repeat his master stroke aboard Cairns' new vessel, the *Ironside*, the world's first iron-hulled sailing ship? Airy, irritated by the havoc caused to his strict working routine by the time spent on the *Rainbow*'s compass, replied in a slightly querulous letter on the "increasing trouble in a matter in which I have no interest." And then, presumably hoping that the large fee would choke off Cairns from accepting his services, suggested a fee of £100 plus expenses to correct the *Ironside*'s compass. (His annual salary as Astronomer Royal was £800.) Much to Airy's astonishment, Cairns accepted under the condition that other owners, if they approached Airy, would be charged the same fee. And would the compass remain corrected for the ship's first voyage to Brazil? Airy assured Cairns that once the compass had been corrected it would be correct for "every part of the earth."

By the afternoon of October 25, Airy was in Liverpool, having traveled north by rail, and was talking in the Royal Hotel with the captain of the *Ironside* about the state of affairs with his new command and the help that Airy had requested: carpenters and three assistants capable of reading compasses and theodolites. Airy also offered, perhaps in a fit of financial remorse at the size of his fee, to correct not only the binnacle compass but also the telltale compass in the captain's cabin.[1]

Even though the ship's binnacle compass was placed 13 feet above the iron of her hull, the captain assured Airy that the compass would require considerable correction. Airy, having heard this, decided to change from one correcting magnet and scroll of iron, as he had used aboard the *Rainbow*, to three magnets without a scroll of iron. But atrocious weather, gales, and slashing rain hampered the swinging of the *Ironside* for Airy's initial measurements, so much so that Airy came close to despair at his decision to travel north. The cloistered calm of Greenwich was calling; and, after a rough correction had been made, Airy quitted Liverpool's driving rain and greasy docks for the sylvan calm of the Royal Observatory perched on the green grass of Croom's Hill overlooking the Thames. But not before leaving with the *Ironside*'s captain most exact written instructions for the final positioning of the magnets.

Crowds of people, drawn by the fascination of seeing the world's first iron-hulled sailing vessel and one whose compasses had been corrected by magnets, cheered and waved as the *Ironside* moved out from Brunswick Dock on the first stage of her voyage to Pernambuco. Two months later Airy heard from Cairns that the *Ironside* had arrived in Brazil and that the compasses had proved correct, the only reservation being that in heavy weather the binnacle compass had proved very unsteady and useless for steering. Airy could only conclude

[1] A telltale, or overhead, compass was hung above the captain's bunk. This way he could check his vessel's course while lying in his berth. Such compasses can be identified, as the East and West points are reversed on the compass card.

that the card was too heavy or the needles too weak. These reservations on the binnacle compass (the telltale compass proved perfectly steady in rough weather) were not mentioned in a short article that appeared in a Liverpool newspaper:

THE IRON SHIP: The first sailing vessel ever built of iron was, it will be recollected, constructed in Liverpool, and was, very appropriately, named the *Ironside*. She sailed for Pernambuco, which she reached after a passage of forty-seven days. Much interest, we may say, indeed, anxiety was felt to know whether the iron would influence the needle. We are happy to state, that the compass was correct throughout the whole passage; and that, therefore, no fear need be entertained as to its general correctness on board of iron-built ocean-going vessels.

Hearing such news made all builders of iron hulls heave a collective sigh of relief. For there had been rumors that Airy's method of correction would not work in seas far distant from where the correction had taken place. John Laird was particularly relieved. He was in the middle of some delicate negotiating with the Secret Committee of the Honourable East India Company to supply two iron-hulled gunboats (Laird called them steam frigates) which would certainly sail in seas far distant from the River Mersey.

The keel of the first gunboat, the *Nemesis*, was laid during the summer of 1839. She was launched in November, ran trials in December, and was ready for sea by the middle of January. She was 173 feet long, measured in at 660 tons, and could float in 6 feet of water when fully loaded with coal, water, provisions, and ship's stores. The stores included ammunition and powder for two pivot-mounted 32-pounders, a number of 6-pounders, and a Congreve rocket launcher. Two 60-horsepower steam engines turned two paddle wheels. Rigged as a topsail schooner, her thin, tall funnel sprouted from her deck just forward of the mainmast. She had watertight bulkheads. As she was flat-bottomed, she had two keels sliding in trunks to stop her slipping sideways when under sail. The shallow draft, to the perspicacious, suggested that this was a gunboat intended for use in shallow waters. And, although it was hinted that she was to sail for Odessa in the Black Sea, the perspicacious would have been right. For she was destined for China's Pearl River and the brewing trouble connected with the seizure by the Chinese of 20,000 chests of opium from British, Indian, French, and American merchants and the virtual imprisonment of the *fan kui*.[2]

The *Nemesis*, after having her compass corrected by magnets using the Airy system, sailed from Liverpool under the fiction of being a "private armed steamer"—even though most of her officers happened to be Royal Navy men. Two days later, in the early hours of the winter morning, she was hard aground on the rocks of St. Ives Bay on the north coast of Cornwall. They were 20 miles off course due to compass error.

Within days the *Nemesis* was in the Royal Dockyard at Portsmouth with workmen crawling over her like ants.[3] This

[2]Hairy barbarians, foreign devils.

[3]The repaired *Nemesis* sailed from Portsmouth on March 28, 1840. According to *The Times*, she sailed "provided with an Admiralty letter of license or letter of marque. If so, it can only be against the Chinese." Months later, after a dramatic and adventurous voyage to the Pearl River, the *Nemesis* was taking a ferociously active part in the 1840–1842 Opium War.

unexpected appearance gave the dockyard's Augustin Creuze an opportunity to inspect her and compare her construction with a wooden ship. He was much impressed with Laird's creation but, in an article published in the influential *United Services Journal*, he opined that: "The effect of the iron upon the compass appears to be one of the most serious objections to these boats." Creuze ended his article by taking an unwise swipe at the Astronomer Royal (a man trained and adept in the vicious infighting endemic in the academic world): "Professor Airey's [*sic*] paper in the Philosophical Transactions is a sealed book to all but men of high science," wrote the innocent Creuze, "and he will not have completed the task he has so ably commenced until he has reduced the subject, both in principles and in its practice, to the level of a far lower standard of knowledge than it requires at present—in fact, to the level of uninstructed common sense."

The Astronomer's irate response to the article was swift and deadly. In a long letter that appeared in the next issue of the *United Services Journal* he pointed out that the *Ironside's* captain had completed the correction of the compass "*before breakfast*" and that Creuze—if he had spent just ten minutes absorbing Airy's instructions as given to the captain on "page 210 of my paper, last paragraph but one"—would also, perhaps, find himself competent to carry out the correction of any compass aboard any iron-hulled ship. That the *Nemesis's* compass was wildly inaccurate had been proved by swinging the gunboat at Portsmouth. As the original correction on the River Mersey had been carried out by a man instructed by Airy, a man in whom he had the utmost confidence, he could only suggest that the compass had been moved. For, "if the compass is too high or low by two or three inches, an error may be produced fully equal to that of the compasses of the *Nemesis*."

Airy's method of correction, refined from that first attempt aboard the *Rainbow*, had been changed to three fixed magnets and iron chain links piled in two boxes. The magnets compensated for the magnetism absorbed by the ship while it was being riveted together on the building slip (known as semicircular deviation) and the chain links for the induced magnetism of horizontal deck beams (known as quadrantal deviation). But deviation was to prove more of a Hydra-headed monster than a dragon. Airy might have been convinced that he had severed two heads (semicircular and quadrantal) with his magnets and chain, but more heads remained—heeling error and sextantal deviation—which were soon to cast doubts on the mechanical method of compass correction.[4]

But what of the reliability of any compass, corrected or uncorrected? Beaufort's energy might have given birth to the Admiralty Compass Committee, but were they achieving anything in the way of an improved compass?

THE FIRST MEETING of the Admiralty Compass Committee took place in the graceful proportions of the Admiralty Library on a warm, humid, cloud-covered day during the summer of 1837. *The Times*, privy to many government meetings, made no mention of it, being more concerned with the tragic death of a certain Mr. Cocking, who had made a fatal parachute descent from a balloon. The Admiralty Compass

---

[4]See the Appendix for more detail on these daunting sounding deviations.

Committee, by contrast, was to prove far more successful than the unfortunate Mr. Cocking.

The six men meeting on that humid day included Admiralty Hydrographer Captain Francis Beaufort and Commander Edward Johnson, as well as Captain James Clark Ross, famous as an Arctic explorer and discoverer of the magnetic north pole, and, it was claimed, also the Royal Navy's handsomest officer; Major Edward Sabine, a friend of Ross who had sailed on two Arctic expeditions as the astronomer and whose particular expertise lay in terrestrial magnetism; Professor Samuel Hunter Christie, a mathematician at the Royal Military Academy and another specialist in magnetism; and Captain Thomas Best Jarvis, a surveyor, late of the Bombay Engineers.

The next three years were spent by various members of the committee inspecting, testing, and analyzing a vast number of British and foreign-made compasses: their needles, cards, pivots, caps, bowls, and gimbals. None of them came even close to the standards that the committee had set for their ideal compass. Their only solution lay in designing, making, and then testing this ideal compass themselves. Prototypes were made and then tested at sea until the committee was satisfied that they had arrived at their goal. The report of their findings, a massive bound volume, was completed on June 29, 1840. (On that date one of the prototype compasses was with Captain James Clark Ross and the 1839–1843 Antarctic expedition with HMS *Erebus* and HMS *Terror*. And Ross remembered his fellow committee members by naming new discoveries of lands and islands after them: Beaufort Island, Mount Sabine, Cape Johnson, and Cape Christie.)

The report asked the Admiralty that twelve more of the

final compass design be made and distributed for further testing. The report specified further that these compasses should
be mounted on a pillar which would be sited aboard the vessel where deviation errors were the least; that these deviations
should be frequently observed by swinging the vessel and a
record should be kept; and that no attempt should be made to
correct the compass by magnets. In short, this compass was to
be considered the standard compass aboard ship and the one
by which the ship was navigated and bearings taken. It was by
this standard compass that the binnacle compass was to be
compared. Thus was born the Admiralty Standard Compass,
Pattern 1, and the Royal Navy's method of correcting for
deviation.[5]

This paragon of a compass was a dry-card compass with its
7½-inch diameter card, elegant in its black-and-white livery,
marked in points and quarter points with a large fleur-de-lis
marking north, pivoting in a black-lacquered copper bowl.
The iridium pivot worked in a sapphire cap. As an azimuth
compass it had an azimuth ring graduated from 0° to 360° in
half-degree increments. It had two vernier scales which could
give readings to one minute of arc, a prism with dark shades,
and a folding sight vane for taking bearings of celestial features. But its most unusual feature was hidden underneath the
card. Instead of the one compass needle it had four needles
and a brass circumferential ring. Thus was eliminated the
worst feature of a single-needle compass: the tendency, so
common in Gowin Knight's compass, for the single heavy
needle to align itself with the direction of the ship's roll.

---

[5]The Admiralty Standard compass was perhaps the longest lived of any
compass, for it was still in use in 1944.

Multiple needles underneath the compass card had been tried before. In 1770 the Danish compass maker Christian Lous had made a compass with two needles and some thirteen years later a compass with four needles. But the problem with multiple needles is that they have to be arranged in a correct pattern. This pattern had been suggested to Major Sabine by a most brilliant and unusual man.

Archibald Smith, the son of a wealthy Glasgow merchant, had followed in Airy's wake and graduated from Trinity College, Cambridge, as Senior Wrangler and First Smith's Prizeman—which meant, to those familiar with Cambridge's medieval traditions, that Smith was a brilliant mathematician. Airy's relaxation happened to be walking. Smith's was sailing. Smith had his own small yacht and was never happier than when cruising along Scotland's West Coast. When the charts were inadequate for poking into secluded anchorages, he made his own. Smith, from his sailing, knew the problems of the marine compass.

It was his suggestion to Sabine that four needles, arranged to a mathematical formula on either side of the pivot, would stop the single needle's habit of swinging when a ship rolled. A few years later he proved, again mathematically, that this arrangement, purely by chance, also corrected for deviations inherent with powerful single-needle compasses when such compasses had been corrected with magnets and iron chain links.

The Admiralty compass was soon adopted by foreign navies, including that of the United States. In 1843, Ross returned to England from the Antarctic. The same year a memorandum was circulating throughout the Royal Navy that raised the Admiralty Standard compass to rank alongside the chronometer as a precious navigating instrument. Both instruments now came under the specific charge of the vessel's commanding officer. Whenever the compass was moved from ship to ship or from ship to shore, it had to be done in the presence of the master or a commissioned officer. Compasses were to be kept in a special compass closet, the key to this closet being held by the master, and detailed instructions were given as to the storage of spare compasses and cards. Matthew Flinders' dream of a Compass Department and an Inspector of Compasses had at last come true, with the driving force behind Flinders' dream being Beaufort. Johnson, Beaufort's fellow committee member, was promoted to captain and appointed to head the new Compass Department. The days when compasses were stuffed, higgledy-piggledy, into the boatswain's storeroom had ended.

The Admiralty's decision to use an uncorrected compass, have their vessels swung for deviation at regular intervals, and keep a record of these deviations in a Deviation Table, stands in stark contrast to the merchant marine, who were leaning toward Airy's mechanical means of compass correction. The battle lines were forming between the two opposing sides. With one side clashing magnets and iron chain links, the other side hoisting standard compasses and tables of deviation, it was only a matter of time before battle broke out between the two opposing forces. A tragic shipwreck in 1854 was the trigger that brought on the conflict.

Chapter 17

THE
"INEXTRICABLE
ENTANGLED WEB"

On a hot February day of 1851, two sweat-caked men on horseback, harnesses creaking and jingling, might have been observed picking a cautious way through the tawny hills and gullies of New South Wales. One of the men, burly and bearded, had spent two fruitless years panning for gold in California but was convinced that gold lay among these Australian hills. At one likely looking creek Edward Hargraves and his companion picketed their horses and started panning for gold from the creek's gravel; four pans out of five produced gold. Hargraves had discovered his El Dorado. "This," he cried to his companion, "is a memorable day in the history of New South Wales. I shall be a baronet, you will be knighted, and my old horse will be stuffed, put in a glass case, and sent to the British Museum."

This happy scenario, alas, never happened. What did happen was the same collective madness that had followed the discovery of gold in California: a massive influx of men and women, stricken with gold fever, seeking their fortune. As the months and years went by, more and more gold fields were discovered. Stories filtered out of miners lighting their cigars with £5 notes, filling horse troughs with champagne, playing skittles with bottles of wine, shoeing horses with gold horseshoes. Ships carrying emigrants from Britain lay abandoned in Melbourne harbor, passengers and crew having headed for the gold fields. Within a decade over half a million people had sailed for Australia. And the building of emigrant ships became a profitable business.

One such ship was the iron-hulled *Tayleur*, hailed as "the perfect emigrant ship . . . which will undoubtedly prove the fastest of the Australian fleet, as she has been constructed expressly with the object of attaining the very highest rate of speed."

When she sailed on her maiden voyage from Liverpool on Thursday, January 19, 1854, bound for Melbourne, she carried 488 passengers in steerage accommodation, 16 in cabin class, 75 crew, and 5 stowaways. The winds at first were light. But within a day it was blowing a gale and the *Tayleur* was soon hammering into a rising sea under reduced canvas. In the early hours of Saturday morning a difference of two points was noted between the steering compass and another compass set close to the mizzenmast. Both compasses had been corrected by Airy's method before leaving from Liverpool. The captain, putting his trust in the steering compass, ordered what he thought was a safe course. Within hours, land appeared under their lee. The ship refused to answer her helm, anchors were dropped, but cables snapped like rotten string.

Within minutes the ship was swept broadside onto the rocks of Lambay Island, about 12 miles north of Dublin. Her maiden voyage had lasted two days.

Lines were flung ashore and a few fortunate souls managed to scramble ashore before the doomed ship slid off the rocks and sank with only her topmasts left showing above the breaking waves. Three hundred and fifty people, most of them women and children, drowned during that dreadful day on the Irish coast. The Board of Trade and the Marine Board of Liverpool inquiry came to the conclusion that the fault lay with the compasses.

Such a tragedy, even in an era when a British ship was lost each day, was a clarion call to those harboring doubts on Airy's method of magnets and chain to correct for compass deviation in iron ships.

WILLIAM SCORESBY HAD abandoned his whaling harpoon in 1823 and taken up the Bible. Souls, not whales, were now his quarry. By the time of the *Tayleur* tragedy Captain William Scoresby had metamorphosed into the Reverend Dr. William Scoresby, D.D., F.R.S., late Vicar of Bradford, now retired and living in the balmier southern air of Torquay. What had not changed was his passion for magnetism and the marine compass. As a founder member in 1831 of the British Association for the Advancement of Science (a Low Church version of the High Church Royal Society), he had been elected to a subcommittee on Mathematical and Physical Science. And over the years Scoresby had been assiduous in attending their meetings and giving readings on magnetism and the compass needle.

Scoresby's magnetic investigations included the nineteenth century's enthusiasm for mesmerism or animal magnetism. Women found Scoresby, with his dark eyes, good looks, and commanding presence, a most attractive man. And Scoresby, in turn, freely admitted being "a high admirer of female beauty." His magnetic experiments for the cure of ailments were conducted with extreme propriety, and the parlors of Torquay soon had obedient ladies lying down on their sofas, heads towards magnetic north, left hands held in Scoresby's right hand (unlike poles attract) as he made left-handed passes over their right sides with a magnet. He found that he had remarkable powers of hypnosis; and one young lady discovered that being mesmerised by Scoresby was a most delicious experience: "My eyes were irresistibly drawn towards him and in vain did I combat against the superior powers of my mesmeriser. A pleasant thrill ran from my finger ends throughout my body towards my feet— my heart bounded with joy and I tasted bliss such as mortals know not. The faces and figures of those around me dissolved, one melting into another until the last vision of them seemed to vanish in Dr. Scoresby's eyes. He was no longer Dr. Scoresby to me, but my all, part of myself—what he wished, I wished."

Such a useful ability, alas, had failed with the Admiralty Compass Committee. Ross had written to Scoresby asking him to furnish compass needles made according to Scoresby's method: laminated hard-steel needles which Scoresby claimed to be far superior in retaining their magnetism than other needles. He had applied for a patent on their construction but assured Ross that he had "no intention of carrying it further if the Admiralty choose to adopt the plan." Scoresby merely asked for public acknowledgment if his compass needles were used by the Admiralty.

Unfortunately for Scoresby, the Committee's man in charge of investigating compass needles happened to be Professor Hunter Christie—a man who held the usual academic's aversion to people they consider trespassers, no matter how gifted, on their territory. Scoresby and Christie had already crossed swords on the subject of compass needles at a British Association meeting. The performance was repeated at a Compass Committee meeting at the Admiralty.

Ross, in a rather churlish letter to Scoresby following the meeting, informed him that laminated needles were nothing new and so no public acknowledgment would be made; also that the Committee (read Hunter Christie) could see no difference between his method of magnetizing needles and that of Dr. Gowin Knight.

No matter; when the Admiralty compass made its appearance, it had multiple laminated needles as suggested by Scoresby.

THE BRITISH ASSOCIATION for the Advancement of Science meeting for 1854 was held in Liverpool. One of the speakers was Scoresby: and his subject caught the attention of *The Times* and *The Athenaeum*. Scoresby's talk combined two emotions always close to the surface of Victorian sensibilities: pathos and righteousness—the pathos of the women and children drowned with the *Tayleur*, and the righteousness that this could have been avoided. Scoresby's talk was entitled "On the Loss of the *Tayleur*, and the changes in the Compass in Iron Ships."

For over twenty years, Scoresby, with thousands of experiments and measurements, had been feeling his way toward a most important characteristic of iron-built ships. All iron ships, he found, had a built-in magnetism—a unique magnetic signature or fingerprint—on leaving their building slip. This was due to the hull's orientation to the magnetic meridian while building and to the hammering, bending, and riveting of her plates. This signature could change at sea under the pounding of waves or the shaking of the hull by engines, paddle wheels, or screw propellers.

A newly launched ship, having been swung and her compass corrected by magnets and iron chain, would sail with her captain placing absolute trust in the corrected compass, but at sea the ship's magnetic signature could change. The fixed magnets could now be a positive danger, malign chunks of steel exerting a wayward influence over the compass. This, claimed Scoresby, is what had happened to the *Tayleur's* compasses—corrected before leaving the Mersey but within two days, after the pounding by seas, showing a difference of two points.

The *Times* reporter noted that such news "has already produced a fluttering among the merchants on 'Change" and that "if the danger pointed out by Dr. Scoresby does exist, neither the shipowners of Liverpool, nor those of other ports, can do any earthly good by blinking the question."

Scoresby's paper brought three reactions. The first was an acrimonious dispute between Airy and Scoresby carried out in a clutch of letters in *The Athenaeum*, where Airy silkily hinted that "when the feelings are excited, the judgement of the speaker as well as his hearers, is very liable to be perverted." He was also convinced that the *Tayleur's* magnetic signature could not have changed in the short time claimed by Scoresby, and that anyone who cared to study the evidence

from two men experienced in Airy's method of compass correction would not be seduced by "the alarmist doctrines of
Dr. Scoresby." The second reaction came from shippers and
underwriters, who refused to have cargoes loaded into iron
ships. The third was the formation of the Liverpool Compass
Committee, funded by local ship-owners.

It was in this atmosphere of utter confusion, with the
building of iron-hulled ships in jeopardy, that the Liverpool
Compass Committee set to work. For this was the nineteenth-
century equivalent of today's jumbo jets falling out of the
skies: something had to be done.

The Committee produced three reports, the final appearing in 1861. Its findings make for terrifying reading when it
lists the cavalier approach of merchant marine officers to the
art of navigation and their compasses. Azimuth compasses
were hardly ever used to take bearings for the measuring
of compass deviation or magnetic variation. Brick dust was
packed into the pivot caps to steady the card. In one case the
vibration of the ship's propeller and the grinding of the brick
dust had made a hole in an agate cap. One captain, on finding
a large deviation of two points in his steering compass,
uprooted the magnet from the deck and moved it closer to
the binnacle, thus doubling the error. Some compass adjusters
claimed that compasses adjusted by them would never show
an error—even though one iron-hulled ship, after having
her compass adjusted with permanent magnets in Liverpool,
showed a dangerous four-and-a-half-point (51 degree) deviation
south of the Cape of Good Hope.

Mainly due to its tireless secretary, W. W. Rundell, the committee managed to steer a difficult course between the conflict-
ing claims of the mechanical and tabular schools of opinion,

making sense of what Captain F. J. O. Evans, head of the Admiralty Compass Department, called the "apparent inextricable entangled web" of compass correction. It was a web shaken by the intellectual feuding of strong-willed men.

One of the committee's recommendations was to site an uncorrected compass aloft as a check upon the steering compass, an idea that Scoresby had been promoting for thirty years. Rundell also recommended that a vertical soft-iron rod be placed near the steering compass to compensate for the induced magnetism when a ship changed her magnetic latitude. Thus, Flinders' idea, the Flinders bar, conceived in the lush landscape of Mauritius, was reborn along the banks of the dun-colored River Mersey.

Archibald Smith had thrown his mathematical weight behind the Admiralty Standard compass and its use with deviation tables. He and Captain Evans, firm friends, were to work together for many years, and between them they created the *Admiralty Manual for Ascertaining and Applying the Deviations of the Compass Caused by the Iron in a Ship*, a manual that was adopted by the United States Navy and translated into French, German, Russian, and Spanish.

The success of the manual lay with Smith's brilliant mathematical analysis of compass deviation due to a ship's magnetism. Both Airy and Smith had based their magnetic analysis on the work of a Frenchman, Siméon-Denis Poisson, who had been drawn to the problem of compass deviation by the Arctic explorations of the Royal Navy in 1818. But his analysis, based on a wooden ship's ironwork, had not included the permanent magnetism of iron hulls. Smith's work allowed for this, and also allowed for two more deviations inherent with Airy's method of magnets and chain: heeling error, and the change

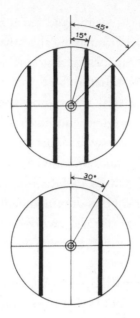

Multiple compass needles as advocated by William Scoresby and Archibald Smith. The illustration shows the arrangement of two or four needles according to Smith's formula.

of magnetic polarity in the ship's vertical soft iron on sailing from the northern magnetic latitudes into the southern. Airy's magnets, in fact, would increase the deviation; and a friend of Smith, Professor William Thomson of Glasgow University, thought Airy's method "most pernicious." The simmering compass war between Smith and Airy came to a boil in 1859.

In 1856, William Scoresby and his wife had sailed to Australia aboard the *Royal Charter*. This was a new 3,000-ton iron-hulled sailing ship with an auxiliary steam engine turning a 14-foot diameter propeller which could be lifted clear of the water when under sail. She had been designed to carry about 500 passengers in three classes of accommodation. She also had a strong room to store gold from the Australian gold fields. Scoresby's object in making this voyage was to prove his point that an iron ship changed its magnetic signature after crossing the magnetic equator.

Before sailing the ship was swung and two of her compasses adjusted with magnets and boxes of chain. She also had an unadjusted compass set 42 feet above the deck on the mizzenmast. The Admiralty provided Scoresby with charts and various instruments which included an azimuth compass, a pocket chronometer, a Fox's dipping needle, and an Admiralty Standard compass—a compass whose needles, as Scoresby wryly noted, copied his principles of needle construction and magnetism.

After sailing, and with a certain amount of tact, Scoresby persuaded the *Royal Charter*'s captain to replace the iron bars of two sheep pens with wooden stanchions, and move bales of hay which had been bound with iron. The pens had been placed close to the steering compass and the hay next to the second compass. The hay was destined for the cattle and sheep, the cattle and sheep for the dining table. Scoresby, perhaps thinking back to the provisions of his whaling days, became rather impressed with the meals eaten at the 50-foot-long dining table. Set with silver dishes, it provided a choice at breakfast of beefsteaks, mutton cutlets, Irish stew, spiced ham, cold beef or mutton, sardines, porridge, freshly baked bread and rolls, and tea or coffee with fresh milk from two cows. The spread for dinner was even more varied: roast beef, roast and boiled mutton, mutton cutlets, mutton curry, mutton pies, roast and boiled chicken, ham, tongue, roast pork, potatoes, carrots, rice, and cabbage—any of which could be tamped down with brandied plum pudding, rice pudding, sago pud-

ding, or fruit tart. Lunch was much like dinner but always
started with soup. Anyone feeling peckish in the afternoon
could have tea, toast, biscuits, and jam.

During the passage to Australia, Scoresby measured the
hull's magnetism and noticed, with some satisfaction, that
the iron hull plates started to change polarity after crossing the
magnetic equator. On the last leg of the passage, running their
easting down among icebergs drifting up from Antarctica, the
hull plates showed even stronger northern polarity, while all
vertical iron (stanchions, an anchor stock standing upright,
and capstans) "had all changed their original magnetism—the
tops now having northern polarity instead of southern."

At Melbourne the *Royal Charter* was swung and her four
compasses compared. The binnacle compass, corrected at Liverpool, now had a maximum deviation of 19¼ degrees and the
other corrected compass a deviation of 17 degrees. The uncorrected Admiralty Standard compass had reduced its maximum
deviation from 25 degrees to 14 degrees. The compass aloft
had hardly changed. The conclusion, as far as Scoresby was
concerned, was obvious. A reference compass aloft was essential for iron-hulled ships. The *Royal Charter* sailed from Melbourne packed with passengers, most of them returning gold
miners, and ten tons of gold and vast amounts of jewelry
packed away in her strong room. Scoresby estimated that the
strong room held close to the equivalent of a million pounds
in sterling.

On his return to England, Scoresby gave a lecture in his
old hometown of Whitby on his voyage:

Every principle I had asserted was completely verified.
The compasses were adjusted on the very ingenious prin-

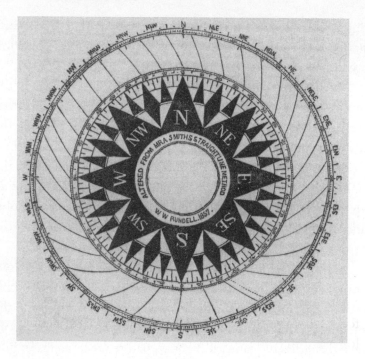

A deviation card. In order to make good a course of WSW magnetic, the *City of Baltimore* would have to steer SW by her compass.

ciple of the Astronomer Royal, the errors being compensated by antagonistic magnets in England. Exactly as I had said before the British Association in 1846, these compasses not exactly ceased to be useful, but they actually went further wrong than any others on board. Every principle of a compass aloft, as the only means of a safe guidance, was fully established. If he cannot combat with an enemy, a wise general gets as far away from him as he can. In our compass aloft we had our perfect guide and standard of reference at all times.

Scoresby died the following year in Torquay on March 21, 1857. But even in death he remained a thorn in Airy's flesh. Scoresby's book on his voyage, *Journal of a Voyage to Australia and Round the World for Magnetical Research*, was published posthumously in 1859. And another thorn, Archibald Smith, had done the editing. But not only editing: Smith had also written a forty-two page introduction on compass deviation which soon made a furious Airy reach for his pen, the result being more column-inches in *The Athenaeum*, a famous public arena for verbal combat. As far as Airy was concerned, Smith should have not used Scoresby's book as a vehicle to proselytize his own views on correcting the marine compass.

The same year as the publication of Scoresby's book the *Royal Charter* was again in the news. On a return voyage from Melbourne, again carrying gold, she was wrecked in hurricane-force winds on the coast of North Wales and 383 people were lost, only 19 being saved from the wreck. However, unlike the wreck of the *Tayleur*, it was not the fault of her compasses.

Chapter 18

GRAY'S BINNACLE

The Commissioners of the 1851 Great Exhibition held in London's Hyde Park had understood the nature of Joseph Paxton's brilliant design for his soaring glass and iron building, for the ground plan was described in ecclesiastical terms: the long central hall (three times the length of St. Paul's Cathedral) was the nave and the cross-hall the transept. The Crystal Palace, as it was dubbed by *Punch*, was a huge cathedral where over six million people, in barely five months, came to worship at the altars of steam and iron. But among the furnaces, steam engines, threshing and hay-making machines, printing presses, looms, heavy artillery, and anchors were wedged more incongruous items: an alarm bed which tilted to tip out the sleeper at the pre-set hour; an alarm clock which not only rang a bell but also, for the hard-of-hearing sluggard, fired a pistol; a knife with three hundred blades upon which thousands of small boys must have goggled with envy; a collapsible piano for a gentleman's yacht; a portable bath with a small furnace attached to provide hot water complete with a solemn warning to the bather: "When the proper temperature is attained, the fire must of course be extinguished."

Charles Dickens twice visited this cornucopia of industry and thought it all too much for him. The young William Morris became physically sick at the overwrought Victorian decoration. Queen Victoria went to it over forty times and got all thrilled and tinglish at every visit, particularly when she considered that her dear, dear Albert had been *instrumental* in organizing the whole extravaganza with its statues, fountains, and giant elm trees growing within the unique structure—a structure that Airy thought would fall down in a gale or shatter in the heat. ("I expressed myself strongly on the faulty principles of its construction.")

Whether Dickens, Morris, Queen Victoria, or Prince Albert noticed the Admiralty Standard compass among the 100,000 exhibits is doubtful. But the compass, Captain F. W. Beechey of the Board of Trade informed the Admiralty Compass Department's Captain Edward Johnson, had been judged the best azimuth compass in the Exhibition.

The Great Exhibition was a supremely self-confident affirmation of Great Britain's industrial might, based on steam and iron, which was sweeping away the old and heralding in the new. Turner's 1838 painting *The Fighting Temeraire*, depicting the wooden relic from Nelson's day being towed away to the breaker's yard by a steam tug, belching smoke and sparks from her tall funnel, paddles thrashing the water, and his rushing vision in the 1844 *Rain, Steam and Speed*, with its dashing railway train emerging from a rain squall, are prime illustrations of this belief.

It's therefore surprising that Britain's navies, the fighting and the merchant, both of them the largest in the world, were obviously sailing on different tacks, marching to different drummers, when it came to the hull construction of their ships.

IN 1827 A French Army officer, Henri-Joseph Paixhans, had produced a new projectile which could be fired from the smooth-bore cannon. It was a round canister, called a shell, filled with gunpowder which exploded on impact, thus neatly turning any wooden hull into a fireship. As protection against this incendiary shell Paixhans suggested that wooden warships be sheathed with iron plates. The ironclad warship had been conceived.

The Paixhans shell was not fired in anger until 1853 when a Russian fleet destroyed the Turkish fleet at Sinope, a port on the Black Sea, followed by the leveling of the port and its fortifications. The result—as if a barrage of these shells had dropped onto their palaces, parliaments, and congress buildings—startled every government with a naval force into the shocked realization that their wooden-hulled ships were in grave danger.

Two years after this dramatic demonstration at Sinope, Great Britain, allied with the French and the Turks, were fighting the Russians on the Crimean coast of the Black Sea. Only a handful of allied ships had steam power and, except for three new French battery ships, floating gun platforms that were used with great success against Russian fortifications, none had the iron cladding suggested by Paixhans.

And what was the reason that Britain's Royal Navy, the world's most powerful naval force, had no ironclads? In the late 1840s the Admiralty had tested the thin iron hull of a rusting harbor-tender with cannon fire using normal solid round shot. The 1850 report on the tests came to one of those magnificently mistaken conclusions: "Iron cannot be beneficially employed as a material for the construction of vessels of war." A panicked Admiralty immediately ordered that eighteen iron-hulled steam frigates, then under construction, be scrapped or converted into troopships. No matter that the *Nemesis* (known as the "Nevermiss") and a sister ship had rampaged with such highly destructive abandon along the Canton and Yangtze rivers during the 1840–1842 Opium War.

"IT IS AN historical fact that the British Navy stubbornly resists change," said Admiral of the Fleet Lord Fisher in 1919. Decades earlier, a similar view had been held by Isambard Kingdom Brunel, the brilliant engineer and designer of the world's first propeller-driven, iron-hulled steamship, the 3,500 ton *Great Britain*. Brunel wrote of the Admiralty's "withering influence" and the "unlimited supply of *negative* principle which seems to absorb and eliminate everything that approaches them."

Brunel's *Great Britain* was christened and launched in 1843 by Prince Albert in a most splendid affair—the Prince Consort traveling down to Bristol from London aboard a special train of the Great Western Railway, driven by its designer, Daniel Gooch, with the company's chief engineer, Brunel, riding on the footplate.

Six hundred people dined in the *Great Britain*'s saloon

before the gates of the building dock were opened and Brunel's creation floated off her building blocks to the din of the cheering crowds, guns firing, church bells ringing, and bands thumping and trumpeting in a patriotic frenzy.[1]

The essence of Brunel is captured in an 1857 photograph where he is shown standing before a massive drum of gigantic chain links, hands thrust into trouser pockets, waistcoat rumpled, stovepipe hat cocked on his head, cigar clamped into the corner of his mouth, and a cool, riverboat-gambler damn-your-eyes look on his face.

In 1846, on her fifth passenger-carrying voyage from Liverpool to New York, the *Great Britain* drove ashore during a pitch-black night on the Irish coast at Dundrum Bay. That no lives were lost is a tribute to Brunel's brilliant use of iron in her construction. A wooden ship would have been totally destroyed. Her captain was cleared of his catastrophic navigation error, but Brunel was convinced that the cause lay with compass deviation.

As more and more iron-hulled merchant ships ploughed across the world's oceans under sail or steam, Britain's Royal Navy was held in the firm and nostalgic grip of Nelson's wooden walls. But the doctrines of Captain Sir William Symonds, the Royal Navy's Surveyor (and also Brunel's arch-enemy), a man who had fought a resolute battle against the adoption of steam power, iron ships, and propellers, were blown out of the water with news from France, news that

---

[1] On July 19, 1970, the 127th anniversary of her launch, the *Great Britain* returned to her dock after a varied career as passenger ship, troopship, emigrant ship, coal and grain carrier, and storage hulk in the Falkland Islands. Restored to her former glory, she is now one of Bristol's prime attractions.

caused much twittering and fluttering in Whitehall's bureau-cratic dovecotes.

It appeared that the French, encouraged by the success of their ironclad battery ships during the Crimean War, were building six ironclad steam frigates. In one stroke the French had made Britain's wooden walls obsolete. The only answer for the British was to build iron vessels belted with thick armor plate which would be faster than and outgun the French *Gloire* and her sister ships. The British riposte to the French, HMS *Warrior*, was launched in 1860, and Captain F. J. O. Evans, the head of the Admiralty Compass Department, was soon grappling with her massive compass deviations caused by thousands of tons of iron.[2]

Since the introduction of the Admiralty Standard compass the Royal Navy had placed its faith in regular swinging to establish and keep track of the ship's magnetic signature, all recorded like a medical record, and from which came the Holy Writ, the table of deviations. But with the *Warrior's* massive deviations, and Evans' finding her "unfortunate compass . . . so tortured in its action," came a mild form of apostasy with the introduction of fixed magnets and a strange double binnacle wherein two compasses (or so Evans believed) would compensate each other. The distance between these two bin-nacles was adjustable so that the correction could be regu-lated. Alas, even though the idea won the Admiralty a gold medal at the International Exhibition of 1862, it proved impractical—and positively dangerous when the *Warrior*

---

[2] The restored *Warrior*, rescued from the ignominy of ending her life as a coal hulk, now floats—burnished and gleaming—as a tourist attraction in Portsmouth Harbour.

changed her magnetic latitude. But the Holy Writ, the Table of Deviations, still remained sacrosanct.

THE COMPASSES OF iron-hulled merchant ships, equally tortured in their action, had led to a new nautical profession of compass adjusters who based *their* creed on Airy's system of compass correction. One of them, John Gray of Liverpool, also made compasses, and in 1854 patented a new binnacle for "Improvements in Adjusting Compasses on board Ships or Vessels." Before Gray's binnacle, all compasses had been adjusted by magnets fixed into the deck. But Gray's binnacle contained correcting magnets which "may be moved in any direction, by screws or rack and pinion, or otherwise, so that when any deviation of the compass is detected, it may at once be remedied." It also included a vertical magnet which compensated for deviations caused when a ship was heeled.

Airy, after his public feud with Scoresby in *The Athenaeum*, decided to go on the offensive. Another paper was read before the Royal Society followed by a letter to the Admiralty pointing out the benefits of his system, which had been adopted by nearly every service in the world except that of Great Britain. Any other system he thought dangerous. And the means by which the Admiralty could fall into line was Gray's binnacle. Surprisingly enough, Their Lordships agreed to a trial of this binnacle aboard an iron paddle wheeler, with the swinging and compass adjusting being made by Gray and Evans. Gray, probably at the urging of Airy, placed two 68-pound iron cannon balls, mounted on brass pillars, on either side of the binnacle to replace the customary boxes of iron chain. The binnacle later won high praise from the captain, but such reports fell on deaf ears at the Admiralty. No matter; Gray, as one of the country's leading compass makers and adjusters, found a more receptive market for his binnacle among a small number of British shipowners and shipbuilders.

A DECADE LATER, however, the Admiralty could congratulate itself on its policy. No Royal Navy ship had been lost to compass error, but the same could not be said of the Merchant Navy. In January of 1866 the readers of *The Times* became aware of the efforts by the Admiralty and the Royal Society to persuade the Board of Trade to regulate the laissez-faire approach of merchant-ship owners and officers when it came to matters of the compass. The Admiralty and Royal Society advocated that every ship should have a standard compass; that mates should be examined on compass matters before obtaining their certificates; that compass adjusters be registered with the Board of Trade; and that a central authority be set up to supervise and enforce decisions made on compass matters.

All this interference, this officious nosey-parkering, met with short shrift from Mr. Thomas Farrer, Secretary to the Board of Trade. "A wide difference," he pointed out, "exists between the relations in which the Admiralty stands to the Royal Navy and the position occupied by the Board of Trade with the Mercantile Marine." Whereas the Admiralty were the owners, designers, and sometimes builders of their ships, the Board of Trade, not owning, designing, or building ships, should not take on the duties of shipowners. "This difference," as Farrer sweetly pointed out, "appears to have been underrated, if not entirely overlooked by the President and Council of the Royal Society." Furthermore, the problems connected with compass deviation—the type of ship, nature and quality of materials, direction on the building slip, type of cargo and its storage, and her position from the magnetic equator—all these were questions on which no authority agreed. The Royal Society might approve of the Admiralty system, but the "totally different principle acted upon in the Mercantile Marine received the support of no less an authority than the Astronomer Royal."

"For these reasons," concluded Farrer, sinking both Admiralty and Royal Society in a final broadside, "the Board of Trade while prepared to procure the best scientific help in investigations into wrecks, are not willing to assume responsibility which would be involved in appointing an officer or officers whose duty it should be to superintend compasses of merchant ships, and to enforce upon shipowners and navigators compliance with what such officers may believe to be the latest requirements of science."

GRAY'S BINNACLE, with its easily adjustable magnets and iron spheres, having eliminated the tedious process of drawing chalk lines on the deck, countersinking the magnets into the deck, bedding them in tallow, and finally making all watertight with a caulked wooden top, meant a significant step forward in compass correction. It predated a similar binnacle by some twenty years.

The conception and birth of this far more famous binnacle came about due to the death of Archibald Smith and, rather surprisingly, a request for an article to fatten the pages of a new magazine.

Chapter 19

THOMSON'S COMPASS
AND BINNACLE

On a September day of 1870, two magnificently whiskered, frock-coated, top-hatted gentlemen rode down on a cinder-strewn train from London to the Solent, and then crossed the dun-colored waters by ferryboat to Cowes. Here they inspected the *Lalla Rookh*, a graceful 126-ton schooner built at Poole in 1854. The forty-seven-year-old Professor Sir William Thomson, limping from an ill-healed broken leg, had brought along his older friend, Archibald Smith, the keen small-boat sailor, to allay Smith's fears that the schooner was far too big a vessel for cruising along Scotland's West Coast.

The two men had much in common. Both had won their mathematical spurs at the universities of Glasgow and Cambridge, both were Fellows of the Royal Society, and both had been honored with medals from that august body. Smith, according to Thomson, if he had not worked as a Chancery barrister, would have been, "with his great mathematical powers and inclination for physical science . . . one of the foremost men of science of this country." Coming from one of the country's most eminent scientists this was praise indeed. Smith's great contribution to safer navigation had been the unraveling of that entangled web of ship's magnetism with a set of coefficients for the numerical analysis of a ship's unique magnetic character and its effect upon the compass.[1]

In 1846, years before this journey to Cowes, when the young Thomson was seeking testimonials for the vacant chair of Natural Philosophy at Glasgow University (a post he obtained and held until 1899), he had been dismayed to hear that Smith also had eyes on the post. The university term ran from November to April, giving a long summer break. In a letter to his sister Smith had waxed lyrical on this aspect of academia: "Then there are six months holidays in the year instead of about *two* and I should . . . get a yacht and make philosophical cruises all summer and live an easy pleasant respectable dull stupid life not toiling and moiling all day long and much of the night as I now do."

This, ironically enough, was exactly what Thomson was aiming to do with the *Lalla Rookh*. Large schooners with a professional captain and crew hint at a wealthy owner. The Ulster-born Thomson, his knighthood stemming from his work on the laying of the Atlantic telegraph cable, had a keen nose for accumulating money through science and its practi-

---

[1]Smith had once written a wry letter to Thomson on the compass fitted aboard Queen Victoria's yacht: "I daresay her majesty little suspects that if the inclination of her compass to the vertical be i there will be a deviation $= \{[f\!\sin\theta-\omega)/f\!\cos\theta]\sin\alpha-[\omega'/f\!\cos\omega]\cos\alpha\}i$."

Sir William Thomson's yacht, the 126-ton schooner *Lalla Rookh*. Sir William tested his compass and sounding machine aboard her. She also served as his summer home.

cal applications. The success of the Atlantic cable lay in large part to inventions patented by Thomson, Cromwell Varley, and Fleeming Jenkin. The three had formed a partnership to share the income from their patented inventions, and Thomson had also formed a partnership with Jenkin to act as consulting engineers to various cable companies. Thomson's income from these sources had made him a very wealthy man, the result being the *Lalla Rookh*, a London townhouse, and the building in Largs of a large, grim fortress of a country house in the then popular Scottish baronial style of architecture.

Two years after the inspection visit to Cowes, Smith was awarded £2,000 by a grateful government for his work on the marine compass. He died, aged fifty-nine, a few months later. For Thomson the death of his friend was a tragic loss. They had known each other since Cambridge days, when Thomson had been a young undergraduate and Smith a Fellow of Trinity College. Over the years they had exchanged views and ideas on ship's magnetism and the compass. Thomson, as a valedictory to his friend, wrote Smith's obituary notice for the Royal Society.

As Thomson composed the obituary notice, compasses were also engaging his attention. In 1871 he had been asked to write an article for the magazine *Good Words*, and had chosen as his topic the marine compass. He then found, rather like Dr. Gowin Knight, much to criticize in its construction. This led him to consider improvements. His first article did not appear until 1874. A second article appeared five years later. "When I tried to write on the mariner's compass," he confessed, "I found that I did not know nearly enough about it. So I had to learn my subject. I *have* been learning it these five years."

Thomson's attention to detail was proverbial. An instance of this can be seen after his purchase of the *Lalla Rookh*, when it came to the choice of cotton or linen covers for the berths: "After anxious consideration and consultation with naval experts," linen covers were finally decided upon, as "the cotton fabric seems to be too hygrometric to be suitable for sea-going places." Best quality damask was chosen for the tablecloths as "drops from the skylight, accidents through want of steadiness of platform, etc., etc., require the strongest resistance against shabbiness of material that the material can give."

After his compass analysis, Thomson arrived at a number of conclusions: the most important one being, following in the wake of Knight, to himself design and make a superior compass, which would be corrected using an improved version of Airy's method based on Smith's years of investigation into ship's magnetism.

Thomson's first attempt, described in a letter to Evans, was shown to the Royal Society of Edinburgh in 1874. Two days later an article appeared in the *Glasgow News*: "Sir William's compass consists of a pair of steel needles, on the model of a needle prepared for the galvanometer by Dr. Joule, the founder of the science of thermodynamics. The needles are each half an inch long, and are supported on a framework of aluminium and glass rods, weighing in all 1½ grains, and hung by a single fibre of unspun silk 1/20 of an inch long." How this compass was to be used at sea was not explained, for it lacked a compass card: Thomson, in his letter to Evans, had objected to "the movable compass card." Evans's reply has been lost, but presumably it was scathing, for Thomson was soon complaining in a letter that the "people in all departments of the Admiralty, except the construction department, are averse to . . . all suggestions from without."

Later that year Thomson had other things on his mind than compass cards and magnetism: for the *Lalla Rookh* made a fast passage to Madeira on a most unusual mission. Thomson's first wife had died in 1870. Three years later, during a cable-laying expedition, Thomson happened to spend some weeks at Funchal and there became acquainted with the Blandys, a family prominent in the Madeira wine trade. The cable ship lay at anchor in the bay and Thomson wrote that "we had admirable lamp signaling several evenings at Funchal between the *Hooper* and Mr. Blandy's house, about 1½ miles distant. The Miss Blandys learned 'Morse' very well and quickly." When the *Hooper* came to sail a figure was seen waving a long white scarf from the Blandy house. "Good-bye, good-bye, Sir William" was the waved message. The sole purpose of the *Lalla Rookh*'s voyage was for Thomson to marry the scarf waver and carry her back to Britain aboard his schooner.

TWO YEARS AFTER this piratical swoop on Madeira, Thomson and his wife were crossing the Atlantic aboard the Cunard passenger liner *Russia*. Thomson had been asked to judge various exhibits at Philadelphia's Centennial International Exhibition of 1876. This was a most convenient voyage, for the *Russia* made an ideal test bed for a compass and binnacle he had just patented. In a letter to a friend Thomson praised its virtues: "We have had a very fine passage across, with just enough rough weather to test thoroughly a new compass, which I shall show you when you come to Glasgow. It behaved perfectly well throughout, notwithstanding a great shaking from the screw (which almost prevents me from being able to write legibly)."

Sir William Thomson's large diameter compass card with its eight needles, thirty-two silk threads, and cut-away card.

This compass bore no resemblance to the one unveiled two years earlier at Edinburgh. Thomson had arrived at his new design after gathering as much information as possible from the Admiralty on its Standard Compass—number and length of needles, weight of cards (the compass came with cards of different weight; a normal card and a heavier card for use in rough seas as it was thought, incorrectly, that a heavy card was steadier in a seaway). Thomson's design incorporated all the features that had been advocated by previous compass designers: a large-diameter card for ease of reading; a light-weight card to reduce friction and wear on the pivot; multiple needles properly spaced to correct for deviation errors and make the card steadier in a seaway; any weight to be at the card's rim.

The design, as he pointed out to a grave and heavily whiskered audience at the Royal United Service Institute, consisted of a 10-inch diameter compass card which looked like no other compass card—except for the degree markings, the compass points, and the fleur-de-lis marking north. The card, braced at its outer edge by an aluminium rim, was cut away in the middle. Thirty-two silk threads, like the spokes of a wheel, stretched from the rim to a small central aluminium boss. A total of eight small needles, 3¼ inches to 2 inches long, were set either side of the central boss. The card pivoted on a sapphire cap set in the boss, the pivot's point being inset with iridium. The whole weight of the card resting on this iridium point came to 180 grains. Compare this, said Thomson, with the 1,500 grains of the 7½-inch-diameter card of the Admiralty compass and the 2,900 grains of the usual 10-inch card used by merchant ships. As to the binnacle, varnished and gleaming, hidden away in its hollow interior were all the correcting mag-

Fig 9.   Fig 10.

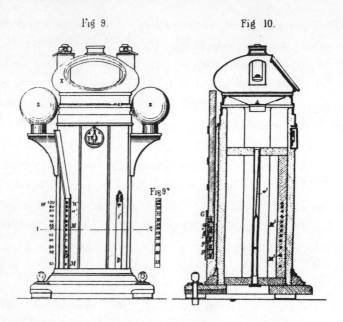

Sir William Thomson's binnacle with its soft-iron spheres on brackets, internal adjustable magnets, and Flinders bar.

nets, laid athwartships and fore-and-aft, all of which could be easily moved and adjusted. Also inside was an adjustable vertical magnet to correct for heeling error, an error which Smith had calculated as producing 2 degrees of deviation for every 1 degree of heel. Outside the binnacle, but attached to it by brackets, were two adjustable soft-iron spheres.

Two years later Thomson was again addressing an audience at the Royal United Service Institute, this time to demonstrate improvements in both compass and binnacle. He had always claimed that his compass was far steadier than other compasses, but when it came to warships and gunfire and steamers vibrating from engines and propellers, the facts were different than his claims. This demonstration was to show that he had grappled with this problem by fitting a new method of suspending the compass bowl in the binnacle. The other improvement was the addition of an adjustable Flinders bar encased in a brass tube.

The Liverpool Compass Committee, years previous to Thomson's binnacle, had suggested the fitting of a soft-iron bar to correct for induced magnetism. Many shipowners had noted this advice and some compass adjusters had followed it with sometimes overwhelming enthusiasm. On one ship the iron bar went down through the deck and into the second mate's bunk. The new binnacle put an end to such inconveniences.

THOMSON'S TIMING FOR the introduction of a new compass and binnacle was masterful. The railway engineer George Stephenson, early in Queen Victoria's reign, had announced his grand plan "to girdle the world with an iron chain, to connect Europe and Asia from their furthest extremities by one colossal railway." Stephenson, like Brunel, had magnificent visions of what could be accomplished with iron and steam. Some thirty years after Stephenson's dream, with the Suez Canal opened and Victoria now a Queen-Empress, the world *was* being girdled with railways and an iron chain made up of passenger ships running, like the railways, to a strict timetable and route. Most of these liners were British-built and British-owned. The journey to India, the great jewel in the Imperial crown, was now measured in weeks rather than months, the vermin-infested horrors of the Overland Route across the Suez isthmus being relegated to anecdotes over the port or brandy. In this apogee of the Pax Britannica a gentleman of means could travel from England to India aboard a vessel of the snobbish and expensive Peninsula and Oriental Steam Navigation Company.[2] Once on board he could start drinking chilled champagne as the liner steamed down Channel, wash down the morning bacon or ham with claret (the P&O were renowned for their clarets and most Englishmen regarded claret as a necessary medicine), and progress through the ensuing days in a happy haze of hock, gin, whisky, brandy, and port (all alcoholic drinks and mineral waters were included in the fare). Three weeks later, having endured the ponderous Sunday services read by the Commander, flirted with the more personable of the young ladies from the fishing fleet,[3] dressed up for amateur theatricals, played deck games, sung in concerts, had serious discussions with pink-cheeked subalterns, grizzled generals, and urbane pro-consuls, he would be decanted at Bombay. Where, if he chose, he could continue on the iron chain and travel by railway to Calcutta or head into the foothills of the Himalayas.

The P&O, in the decade after the opening of the Suez Canal, built over two dozen vessels to steam on a regular basis from Britain to India. The P&O's rival, the British India Steam Navigation Company, famous for its cockroaches, was also engaged in a fury of shipbuilding. On the North American

[2]This famous line, the P&O, was also known as the "Dear and Slow" and immortalized by Rudyard Kipling in his *Exiles' Line*, a parody in the style of Edward Fitzgerald's *Rubaiyat of Omar Khayyam*. The P&O had a high regard for itself: no Board of Directors for this company that carried the Royal Mail to India and Australia; a Court of Directors, please, and remember that all our Captains are Commanders, and all our shareholders are proprietors.

[3]Young ladies bound to India for no other purpose than to catch an eligible, preferably rich, husband.

route two competing lines, the Cunard and the White Star, were trying to outdo each other with larger and faster ships. South Africa was served by two rival companies, the Union Steamship and the Castle Packets companies, also building larger and faster ships.

This worldwide web of steam and iron also had a growing web of copper in the telegraph cables that spanned continents on poles and lay hidden in the ocean deeps. Now, through the chatter of the Morse machines, shipowners could talk to their agents and orchestrate the movement of their ships to take advantage of the market and load and unload cargoes. For this type of traffic a new ship was born, a workhorse of the oceans, the tramp steamer.

This massive increase in British shipbuilding by the mercantile marine was matched by Britain's Royal Navy. Warships, small and large, were being built to counter any threat from France, Germany, or Russia. And these warships, their names suitably belligerent—*Hotspur, Hercules, Audacious, Devastation, Colossus, Conqueror, Thunderer, Inflexible, Invincible*—all with spotless paintwork and gleaming copper and brass, their White Ensigns fluttering in the sea breezes, steamed out from their naval bases that encircled the globe—Bermuda, Simonstown, Halifax, Esquimault, Gibralter, Malta, Cyprus, Hong Kong, Aden, Trincomalee—intent upon keeping the Pax Britannica.

All this vast tonnage of British shipping, as well as the rest of the world's shipping, required reliable compasses to battle with the demon of deviation. Thomson was determined that they should be his.

Chapter 20

THE SELLING OF
A COMPASS

The *Lalla Rookh* proved an ideal vessel for Thomson's testing of his prototype compass. The copper-bottomed schooner, with her gleaming brass, glossy varnish, scrubbed decks, linen bedding and damask tablecloths, also made an ideal means of admittance into the world of yachting, a recreation which, except for some notable eccentrics, was mainly one for the wealthy and the artistocratic.

Thomson, for two decades, spent his summers aboard the schooner. After a winter lay-up in the Gareloch, the sailing season would start with a shake-down cruise in the Firth of Clyde, followed by the passage south to the summer base at Cowes. From there, Thomson would make cruises along England's south coast and abroad. The season ended with a September cruise along Scotland's West Coast.

Some of Thomson's guests were less than enthusiastic about the joys of yachting. One disenchanted fellow morosely observed that he would have enjoyed it still more if the schooner had never left the slip, another that the "best thing about yachting was going on shore."

One man who endured it all with stoic Prussian fortitude was Professor Hermann von Helmholtz. The world-famous physicist, invited to join the *Lalla Rookh* at Inverary, where forty other yachts were gathered, along with a host of Campbells, to celebrate the marriage of the Duke of Argyll's son to Princess Louise, Queen Victoria's daughter, wrote letters to his wife on his Scottish experiences. From St. Andrews, before joining the *Lalla Rookh*, he described golf as "a sort of ball game in which the ball lies on the ground and is continuously struck by special clubs until it is driven, with the fewest possible blows, into a hole, marked by a flag." It was a game, he wrote, which was played "with great vehemence." After his introduction to golf the bemused Helmholtz related his first experience of yachting:

My cabin is just about so large that I can stand upright in it beside the narrow bed: the rest of the space is less lofty, yet it contains wash-table, dressing table, and three drawers, so that I can arrange my things well. For washing the space is rather small, particularly when the ship rolls and one cannot stand firm. To-day we began the morning by running on deck wrapped in a plaid and sprang straight from bed into the water. After that an abundant breakfast was very pleasant. Then came visits to the other yachts, and so the day has up to now passed very pleasantly in spite of the rain.

All this was yachting experienced at anchor. With the schooner under sail, Helmholtz reeled around the deck trying to keep his balance and having "cataracts of sea-water . . . run off my waterproof."

Yachting brought Thomson into the orbits of the country's power brokers. One such man was Lord Dufferin,[1] who also happened to be chairman of the Admiralty committee looking into the catastrophic sinking of HMS *Captain*, a new vessel of unusual design which had capsized during a gale in the Bay of Biscay with close to 500 men drowned. The majority of the committee were experienced naval officers, with a sprinkling of civilians, Sir William Thomson being one of them. The committee met in London every two weeks for several months. They also visited dockyards, were escorted around warships, drank and ate with naval officers. These contacts were to play an important part in Thomson's marketing campaign for his compass and binnacle.

If his summers were spent cultivating naval officers in the south, Thomson's winters were spent in the north giving his morning lecture at the university, followed in the afternoon by a visit to James White's workshop to inspect the manufacture of his patented instruments, which included the compass and binnacle.

Glasgow, of all the world's cities, was the ideal marketplace for the selling of nautical instruments to shipbuilders and shipowners. In 1876, the year of Thomson's patent for the compass and binnacle, yards on the Clyde were building more

---

[1]Dufferin had sailed his schooner, the *Foam*, to Iceland, Jan Mayen, Spitzbergen, and Norway in 1852. He was later Governor-General of Canada and Viceroy of India. In 1872, with Earl Ducie as seconder, he proposed Thomson for membership in the Royal Yacht Squadron.

iron ships than the rest of the world put together; and from then on, until 1914, the Clyde launched one-third of all the new tonnage built in Britain. Across the Atlantic, in the United States, wood was still the main shipbuilding material. Any American iron hull builder or steam engine manufacturer came face to face with the American ironworks and their preference to roll and supply iron for the building of railway tracks and bridges. None of them wanted to roll iron plates for ships. Even propeller shafts had to be ordered from abroad.

In 1876, the same year that Sir William and Lady Thomson had crossed the Atlantic aboard the *Russia*, the first commercial Thomson compass and binnacle was fitted on an iron-hulled steamer running between Glasgow and Belfast. A few months later, compasses and binnacles were being fitted aboard the White Star liners *Britannic* and *Germanic*, vessels of 5,000 tons and, at 15½ knots, the fastest vessels on the Trans-Atlantic run. The P&O soon followed the White Star's example. The British India proved a harder nut to crack. The captain of the *Chanda*, after his ship's maiden voyage to Calcutta, complained that the compass, with its adjustable magnets, was too complicated and delicate an instrument for simple seamen and the rigors of the sea. But the captain's reservations were soon swept aside before the sales onslaught of Thomson and his agents, and the British India fell into line behind the P&O. Other captains from other lines also voiced doubts, finding the compass unsteady in a seaway. One company even went so far as to request Thomson to remove the compasses and refund the purchase price.

One man firmly on Thomson's side was a fellow Ulsterman, the improbably named Squire Thornton Stratford Lecky, a captain in the merchant service and author of *Wrinkles in Practical Navigation*, an influential book first published in 1881 that went through twenty editions in forty years. Lecky was to perform the same service for Thomson as John Robertson had with his *Elements of Navigation* for Dr. Gowin Knight. "When the owner's pocket can afford it," wrote Lecky, "there is no better Standard compass which in any way can rival the one invented and patented by Sir W. Thomson, of Glasgow. Its mechanical construction is as near perfection as may be; and looking at it either theoretically or practically, it has advantages which no other known compass possesses."

For important commissions, such as the Tsar Alexander's yacht the *Livadia*, Thomson himself would visit the vessel and adjust the magnets for compass correction. Today, looking at a Thomson binnacle in a maritime museum, it is easy to conjure up the bearded figure crouched over his creation, delving into its innards like some mad scientist from an early Hollywood epic. Here, with its compass bowl topped by a glass-fronted dome (head), oil lamps (ears), iron spheres on brackets (vestigial arms), octagonal varnished binnacle (body), and protrusions at base for fastening to deck (feet), is a humanoid-looking robot, being readied to take over the world.

In 1876 the factory of James White, a Glasgow instrument maker, had made a mere dozen compasses. A decade later, with Thomson now the major partner, the company had moved to larger premises, employed two hundred people, and was producing over three hundred compasses a year. By 1892, with Thomson now signing his letters as Kelvin (having been raised to the peerage as Baron Kelvin of Largs), the company was producing over five hundred compasses a year. In 1900 the business was formed into a limited company and became

Kelvin and James White, Ltd.[2] On Kelvin's death in 1907, over ten thousand compasses and binnacles had been made in the Glasgow factory.

The Thomson compass and binnacle, just like the Knight compass, was an expensive item. And Thomson, just like Knight, used every opportunity to promote his design with lectures, articles, and letters to men of influence. One such letter went to the First Lord of the Admiralty, W. H. Smith,[3] requesting that the compass be introduced into the Royal Navy. He enclosed a list of sixty large steamers and sailing ships that were using his compass, plus the many reports that, "after eighteen months of very varied experience in all seas and all weathers," proved the superiority of his design over any other "compass hitherto in use." As a marked hint that the Royal Navy was lagging behind the navies of other countries, Thomson went on to point out that the Germans had given the compass a six-month trial on the ironclad *Deutschland* and had ordered a second compass. The Russian, Italian, and Brazilian navies had also ordered compasses.

Thomson then continued, with a certain amount of bland cheek, that "from results of trials on board ships of war firing heavy guns I have recently made some improvements in my compass." Thomson had taken the unorthodox approach of supplying his compass to amenable captains in the Royal

---

[2]More name changes followed and in 1947 the company amalgamated with Henry Hughes & Son, the London instrument makers, to become the present-day Kelvin Hughes.

[3]This was a political appointment. W. H. Smith (1825–1891) was a newsagent, bookseller, and politician. In 1849, he secured the privilege of selling books and newspapers at railway stations. Today the company is one of Britain's largest newsagents and booksellers.

Navy. The trials aboard Her Majesty's ships had been unofficial, with the captains making their reports not to the Admiralty, but to Thomson. One of the captains, on being questioned as to why he had fitted the compass without obtaining authority, airily replied that as it "was lent by a mutual friend for private purposes . . . no authority being required, none was asked for." Thomson's cultivation of naval officers during the *Captain* inquiry, the summer cruises aboard the *Lalla Rookh* entertaining officers, and cruises aboard Royal Navy ships as the captain's guest were bearing fruit.

The Admiralty, of course, had strict rules as to how their ships were to be navigated, and the central core of the method was that officers should never assume their compasses to be correct. Constant attention in measuring for deviation was essential. Any other approach was a virtual thumbing of the nose at the Admiralty and its directives.

But Thomson's compass, according to its inventor, had all "the qualities necessary for thoroughly satisfactory working in all weathers and all seas, and in every class of ship." Plus, with the binnacle containing the magnets and soft iron, "correctors can be applied so that the compass shall point correctly on all points, and those correctors can be easily and surely adjusted at sea, from time to time, according to the changing position of the ship."

Such a statement was sweet and seductive music to the ears of some officers, both merchant and naval, who found applying compass variation and compass deviation to be a headache-inducing, bewildering navigational problem. Here, according to Thomson, was the compass Holy Grail. Gone was all the confusion of applying deviation errors and repeated swinging.

If Thomson had Lecky as his apostle in the merchant marine he also had one in the Royal Navy: Captain "Jacky" Fisher.

FISHER, LATER LORD Fisher of Kilverstone and Admiral of the Fleet, was a whirling tornado of a man who is remembered for his sweeping reforms of the Royal Navy and the introduction of HMS *Dreadnought*, the first of the big-gun battleships. He had joined the Royal Navy in 1854, aged thirteen: his entrance examination consisting of writing out the Lord's Prayer, reciting the three times table, and jumping over a chair in the presence of a doctor. He was then given a glass of sherry by his gold-epauletted, blue-coated, whiskered examiners, as evidence of his becoming a naval officer, and entered into the logbook of HMS *Victory*: "Joined Mr. John Arbuthnot Fisher."

Fisher's great passions were his country, the Navy, dancing, and listening to sermons. *Speed* is the one word that comes to mind when thinking about Fisher. All ships, from launches to battleships, had to be driven at full speed. Even clearing away the tablecloths, cutlery, and plate after an *al fresco* dinner on the quarter-deck had to be done at speed—three and a half minutes being allowed, and damn the breakages. Fisher's letters, full of exclamation marks, capital letters, and vigorous underlining, give a sense of this manic energy.

The late Victorian Navy on which Fisher cut his teeth was a Navy where spit-and-polish was more important than gunnery practice—it made such a *fearful* mess of the paintwork—and the exercises and practices cut into an officer's dashing social life. Steam engines were looked on with distaste by deck officers as coaling was such a messy business.

When, in 1879, Fisher was appointed Flag-Captain to the newly built *Northampton*, he took command of a curious mixture of antique Nelsonian Navy stirred in with the latest in Victorian technology. She was ship-rigged and could cruise under sail or steam; she had twin propellers; searchlights; torpedo tubes; telephones; Nordenfelt guns. After commissioning she went on a week of trials in the English Channel: trials under sail (where it was found she carried 15 degrees of weather helm and could only wallow along at 3 knots in a good sailing breeze); trials under steam; trials under sail and steam; sail handling; gun trials; general quarters; night quarters, searchlight tests; coaling ship.

Those were the days when every Briton had an opinion, no matter how ill informed, on naval matters. England's wooden walls might have been replaced by iron and steel walls, but the Navy's duties stretched around the globe. Naval correspondents held a unique position in the daily press and as the *Northampton* was a new and formidable addition to the fleet, *The Times* devoted a complete article to her trials—noting that Sir William Thomson was on board. His compass, although the *Northampton* had her usual Admiralty Standard compasses, had been put on board at the request of Fisher.

The Times reported on the compass in glowing terms: "Sir William Thomson's inventions were subjected to very practical tests, and were found to answer perfectly. During the gunnery practice the vibration of the compass was inappreciable, even when firing broadsides. . . . On the other hand, blank charge fired from a 20-pounder caused the card of the Admiralty compass to oscillate violently; and it would have been impossible at any time during the target practice, when the firing was more or less continuous, to steer except by the *untrustworthy method of using a liquid compass.*"4 Thomson promptly had 500 copies of the article reprinted and distributed to merchant ship owners.

Fisher now became Thomson's fervent apostle within the Admiralty, and also in the public arena. Thomson, in order to protect his patent, had brought a series of court cases against other compass manufacturers; Fisher and other naval officers acted as witnesses on the excellence of the Thomson compass compared to the Admiralty compass and other compasses, although one officer did admit that during the 1882 bombardment of Alexandria his ship had been steered, not by a Thomson compass, but by a liquid compass.

Fisher's method, within the service and out in public, was to ridicule the Admiralty compass, and in a series of letters to Thomson the ardent Fisher reported on his campaign to have his compass replace the Admiralty compass. When in London Thomson would stay with Fisher. By 1889 the two men had won their campaign. This was a major financial triumph for Thomson. For, as Fisher wrote to him, the Admiralty would no longer make the old compass: "But twenty of yours to be kept in stock and all ships to be fitted with two, one for steering and one for taking bearings, but please do not mention anything about this." But Thomson, now Lord Kelvin, was not content with this. He would visit the Deptford compass store, examine the stock, and then send the Admiralty an order list to be sent to Glasgow.

Thomson's selling of his compass had been a carbon copy of Knight's method: the cultivation of powerful men; the demonstrations and public lectures; articles in influential jour-

4The italics are mine.

nals; the puffery in books and newspapers. Ironically, just as the Knight compass had critics pointing out its unsteadiness in a seaway, so did the Thomson compass have its detractors making the same point: its unsteadiness in a seaway, under gunfire, and from the vibration of ships being driven at speed.

One of Fisher's biographers, Admiral Sir Reginald Bacon (a fully paid-up member of the "Fishpond"),[5] thought that Fisher "never made a mistake as far as the Navy is concerned." But, with that driving energy in the cause of the Thomson compass, Fisher did make a mistake. Not all naval officers were happy with the compass: Captain Ettrick Creak, head of the Admiralty Compass Department, in particular. After his retirement in 1901, Creak wrote to the Admiralty Hydrographer: "When the Thomson Compass was first introduced as the Standard Compass on board I felt it my duty to try and make it a success. It was, however, in many respects the *bête noire* of my existence."

CREAK'S *BÊTE NOIRE* ended its Admiralty life in 1906. Its replacement was a new type of compass, a liquid compass, which Creak and another naval officer, Commander L. W. P. Chetwynd, had been perfecting for decades. Admiral Sir Mostyn Field, years later, was to write that "Creak had to patiently face great and ignorant obstruction—official and other—before he succeeded in getting his compass installed." The Royal Navy, in fact, with its adoption of the Thomson

[5]The faithful band of officers committed to Fisher and his naval reforms. Anyone outside the group stood small chance of any important naval command. Bacon, the Fisher acolyte, received the plum command of the world's mightiest battleship, HMS *Dreadnought*.

compass, was falling behind other navies who had long before changed over to liquid compasses.

For Kelvin, the usurpation of his dry-card compass by a liquid compass was a blow to his pride, pocket, and ego (he was firmly convinced that all other compasses were inferior to his). In a letter to his agent, after hearing the ominous news that the liquid compass was now king, he voiced his worries: "This Admiralty business is very serious. It seems they have quite resolved to have Captain Chetwynd's liquid compass all through the Navy and to displace mine everywhere. We have had no Admiralty orders for compasses during the last two or three years. They have about 3000 of my binnacles lying unused in dockyards, some of them not taken out of the cases in which they came from us."[6]

1876, the year that Thomson patented his first compass and binnacle, was also a year that contained the nemesis for his new compass. One of his reports on the instruments at the Philadelphia Centennial Exhibition had been Report No. 230: *Ritchie's Floating Compass*, a compass which had been chosen by the United States Navy as its standard compass and so, as one authority has succinctly put it, placed that navy forty years ahead of the Royal Navy when it came to compass matters.

Another portent for the death of the marine dry-card compass, for those with the wit to see it, was the launching in 1877 of a small vessel that looked more like a River Thames steam launch than a warship. HMS *Lightning*, 85 feet long, 11 feet wide, and with a speed of 18 knots, was to shock the world's navies into acknowledging the somber fact that their

---

[6] At about £50 a binnacle, this represented a massive amount of money.

cherished battleships were in grave danger. For the innocent-looking *Lightning* was the world's first torpedo boat, and carried the newly invented, and quite destructive, Whitehead self-propelled torpedo packed with high explosives. Sea tests of this new breed of high-speed craft showed that the dry-card compass, compared to the liquid compass, was utterly useless.

Chapter 21

A QUESTION OF
LIQUIDITY

The launching of the *Lightning* from Thornycroft's
yard on the River Thames soon spawned a breed
of larger and faster torpedo boats. By 1890, every
self-respecting country with a navy had to have a flotilla of
these rakish and dangerous craft. France led the world with
210, closely followed by Britain with 206, and the navies of
Germany, Italy, Russia, Austria, Greece, Holland, Denmark,
China, Norway, Sweden, Turkey, Japan, Brazil, Argentina, and
the United States added another 825.

But no matter what flag fluttered from their ensign staffs,
they all shared the same characteristics. They were hellishly
uncomfortable and very fast: the fastest boats in the world.
Thundering along in a seaway, their funnels streaming smoke
whipped horizontal by their speed, ensigns and pennants
thrashing in the wind, throwing huge bow waves, knifing

through wave tops and thundering down into the troughs, they were warships for young men. And the young men loved them. For they were the Davids, flinging their torpedoes from deck-mounted tubes, who could slay the ponderous and slow-moving battleship Goliaths.[1]

The British Royal Navy painted their piratical craft black, which only added to their menace and sense of pure utility. Spit and polish played no part of torpedo-boat life. The men, ratings and officers alike, dressed in thick serge suits and seaboots. In cold weather, they wore hooded coats made from duffel. Washing, water being a precious item, was low priority. All this, plus extra pay and an undiluted rum issue, led to a swashbuckling esprit-de-corps among these nautical Davids.

Having uncorked this demon that now menaced their own mighty battleships, and with a wary eye cocked on the French, whose torpedo boats lay like wasp nests just across the Channel, the Admiralty was now presented with an exquisitely embarrassing problem. And so, to slay this demon, larger vessels were built, torpedo-boat catchers, to sink these pesky vessels. But they proved slower than their prey and thus an expensive failure. The answer to this dilemma came from Alfred Yarrow, a builder of torpedo boats at Poplar on the River Thames, who approached Admiral Fisher, now Controller of the Navy and responsible for the design and building of the Navy's ships. Yarrow said that he had the details of France's latest breed of torpedo boat and could guarantee to

[1]The first torpedo launched above water was from HMS *Actaeon* in 1875. It was launched in a truly amateur and English style. A tilted mess table with the torpedo resting on top was placed by an open scuttle. When all was ready the torpedo, propeller whirring, was pushed out into the water to a chorus of loud huzzahs.

build faster and better ones. Fisher jumped at the chance. Two vessels were launched, 180 feet long, 18 feet wide, with engines developing 4,000 horsepower. HMS *Havoc* and *Hornet*, known as torpedo-boat destroyers, speeding along at 28 knots, became the world's fastest ships. Fisher was enchanted and ordered thirty-six more to be built by yards specializing in the building of torpedo boats.[2] Within a few years the British had a destroyer flotilla of ninety ships. The Admiralty, seen with a jaundiced eye, at times might stumble and bumble, but, over the centuries, it has never lost a trick with the naming of their ships. The destroyers' names reeked with their dashing purpose: *Daring, Ferret, Lynx, Banshee, Dragon, Rocket, Shark, Ardent, Boxer, Bruiser, Spitfire, Swordfish, Charger, Dasher, Fervent, Conflict, Hasty, Desperate, Thrasher, Hunter, Vulture, Sparrowhawk, Virago.*

Torpedo boats and destroyers shared the same characteristics: they were uncomfortable, very wet, and, when pounding along at full speed, vibrated in such a paroxysm of energy that rivets were often fired from their plating. Steering with a dry-card compass, such as Thomson's or the old Admiralty Standard compass, was impossible. All this was proved in 1881 in some rather lively sea trials aboard a torpedo boat fitted with a liquid compass and Sir William Thomson present with his dry-card compass. The liquid compass won outright, Thomson's compass being pronounced "quite unserviceable." Three years later, Thomson was back again demanding more trials with what he claimed was an improved compass. The results were the same, the report noting that the liquid compass han-

[2]Fisher had asked Yarrow what his design should be called. "That's your job," replied Yarrow. "Well," replied Fisher, "we'll call them Destroyers, as they're meant to destroy the French boats." The name stuck.

dled "the lively movements of the boat in a seaway when Sir William Thomson's fails under the same conditions."

A few years later, in 1890, four compass makers were requested by the Admiralty to submit designs for a liquid compass to be used aboard small steamboats and in boats under sail or oars. A compass by E. J. Dent, who had been making liquid compasses for the Navy for over fifty years, was finally selected. This decision brought a prompt query from the Controller of the Navy. Had Sir William Thomson been invited to submit his compass? No, was the reply: for the simple reason that dry-card compasses had been considered unsuitable for small boats since the 1850s. No matter; a letter was promptly sent off to Thomson asking him to submit a compass to enter the lists against the Dent compass. The Admiralty also offered Thomson all their facilities for making trials and experiments at Portsmouth: a convenience which had not been offered to the other compass makers. One year later, Thomson's dry-card compass was tested against the liquid compass. It was found to vibrate through eight points while the liquid compass remained almost steady.

This farce, delaying the introduction of Dent's compass by a year, illustrates the long shadow cast by Thomson and what one officer called "individuals often pleasant, persuasive, persistent, and perceptive of everything but possible faults in their inventions." A few years later, in 1896, with the publication of a Royal Navy handbook, *Torpedoes & Torpedo Vessels*, the last rites for the dry-card compass were pronounced: "The boat is nearly always steered from the wheel outside the forward conning tower. Liquid compasses are generally used, the vibration and motion being too violent for the delicate instruments invented by Sir William Thomson."

. . .

THE LIQUID COMPASS, wherein the compass needle is pivoted in a liquid-filled bowl, was not new. In 1588, a few weeks after the defeat of the Spanish Armada, the Suffolk-born Thomas Cavendish, following in the wake of Sir Francis Drake, returned from his plundering expedition around the world. The *Desire* and her crew, loaded with booty from the Manila galleon and anchored off Greenwich, were given a most gracious reception from Queen Elizabeth and her court. And a morose Spanish agent noted some of the Queen's caustic and amused comments: "The King of Spain barks a good deal but does not bite. We care nothing for the Spaniards; their ships, loaded with silver and gold from the Indies, come hither after all."

Aboard the *Desire*, and feeling the chill of the November air, were two sailors from the East Indies. William Barlowe, not yet an archdeacon but starting on his avocation to improve navigation, eagerly questioned them on compasses used in their seas. They told him that they used a long needle, about six inches long, pivoted in a bowl of water. It had no compass card, just two lines at right angles to each other painted on the bottom of the bowl. It's a pity that Barlowe, an ingenious man who created many instruments to ease the problems of navigation, just recorded the fact and took the idea no further.

Some two hundred years later the Dutch-born Dr. John Ingenhousz, a large and cheerful medical man who had made his home in England to study methods of smallpox vaccination (and who also made rather alarming experiments with inflammable gas; ignited candles with an electric spark; collected gas from cabbage leaves, keeping it in bottles which were then stuffed into his capacious pockets), gave a talk at the

Royal Society suggesting methods of steadying the compass needle. All used a glass-topped, liquid-filled bowl with the needle pivoted on a vertical pin. To reduce friction on the pivot, the needle either had a float made from cork, or was enclosed in a glass tube.

A few years later, Gabriel Wright made a liquid compass very similar to that described by Barlowe. It had a lacquered needle to prevent rust which pivoted in a water-filled bowl. It lacked a compass card, but the compass points were painted on the bowl's bottom, the needle acting like the pointer of a dial. Tests by the Royal Navy found it to be far steadier than the dry-card compass, particularly in small boats with their quick motion in a seaway. But the Navy failed to develop the idea.

In 1813 a massive advance was made in the design of the liquid compass when Francis Crow, a watchmaker and silversmith of Faversham in Kent, was awarded Patent No. 3644 for "Certain Improvements in the Mariner's Compass or Boat Compass." It was a brilliant design that contained many of the features common in the modern liquid compass.

To eliminate the problems of freezing, the glass-topped and black-lacquered bowl held spirits of wine rather than water. It also had an expansion chamber to allow for the liquid's expansion and contraction. But the most important feature was the hollow circular float (Crow called it a lens) painted with the points of the compass and made from copper that formed the compass card. The compass needle was sealed within this float. On the top surface of the float was cemented an inverted cone and a weight on a spindle was cemented on the underside. Cemented to the underside of the bowl's glass top was a pivot point that aligned with the cone. The float had sufficient buoyancy to rise against the pivot with a total

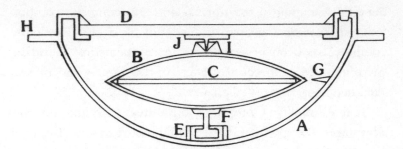

Francis Crow's patented liquid compass of 1813. This patented compass predated the American liquid compass of E. S. Ritchie by fifty years. Ritchie's compass also used a pivot point on top of the float but substituted another pivot in place of Crow's weight. The following description is taken from Crow's patent:

A Copper bowl filled with alcohol.

B Float or lens made from copper and painted on top with the points of the compass.

C Magnetic needle.

D Thick glass top.

E A ring of copper to prevent float from being thrown from the point or center of action.

F A weight to keep the float in a horizontal position and to adjust the pressure on the point of action to about twenty-four grains.

G The lubber's point.

H The arm or point of suspension coming from the bowl which is supported by the gimbal ring in the usual way.

I Inverted hollow cone.

J The point of action which is riveted on a copper plate and cemented to the interior surface of the glass top.

weight of about "twenty-four grains troy weight, whereas the weight acting on the point of suspension in the common mariner's compass is seldom less than one ounce and a half, or seven hundred and twenty grains." The liquid also dampened the wild swinging of the normal dry-card compass and thus reduced wear on the pivot point. And, concluded Crow, his compass "has been proved to retain its magnetical meridian when exposed to the utmost effect of the most turbulent sea, even in a boat of the smallest size."

The Admiralty Compass Committee had ignored the advantages shown by Crow's liquid compass—or, for that matter, any other type of liquid compass. But by 1845 the Admiralty had found it necessary to issue liquid compasses for use in rough weather, for it was obvious that any liquid compass was far steadier in a seaway than a dry-card compass.

Francis Beaufort, in his position as Hydrographer, was well aware of the importance of accurate compasses for survey work and urged that comparative tests be made between the two types of compass. In 1847, aboard the steamer *Garland*, tests were made between six compasses: three dry-card compasses and three liquid compasses. The tests were made in the North Sea on passages between Dover and Ostend. The outward-bound passage to Ostend proved calm, and all the compasses performed admirably. The return passage, however, was very different. Gale-force winds, strong currents, and the North Sea's shoal waters churned up what one observer called a "kind of wrangling sea" and one where the brand-new Admiralty Standard compass, on one particularly vicious stretch of sea, swung through 12 points and one of the other dry-card compasses through 16 points. All three liquid compasses proved far steadier. Further tests, this time to measure the effects of gunfire, reinforced the superiority of the liquid compass.

But the Admiralty had invested too much time and money in its Standard compass for it to be superseded. And so the

Royal Navy's ships in bad weather either replaced the normal card with a heavier card (which only made the swinging worse), or replaced the compass with a liquid compass.

Other navies were not so wedded to the dry-card compass. In 1862, E. S. Ritchie, an American inventor living in Brookline, Massachusetts, designed a most unusual compass, rather like a periscope, for the ironclad USS *Monitor*.[3] The compass, designed to be free from the deviation errors caused by the *Monitor*'s armor-plate, although placed outside the steel steering tower on a tall and thin binnacle, could be read inside the tower at the steering position. Later that year, Ritchie patented a liquid compass similar to Crow's in that it had a float housing the compass needle. U.S. Patent No. 36,422, "Improvement in Mariners' Compass," was the first of many compass patents taken out by Ritchie.[4] One of his later patents was for the paint to be used on the compass card. Paint had always been a problem for liquid compasses where the liquid was alcohol or a mixture of alcohol and water. Ritchie's patent was for a paint made up from dry white-lead and egg white painted on the card, and then hardened by a solution of lime or by heating. By the early 1870s the United States Navy had adopted the Ritchie compass and could congratulate themselves that the "United States Navy is possessed of a compass unsurpassed in its essential qualities by those in other services whether naval or mercantile."

[3]The *Monitor* has entered naval history. A dangerously low-freeboard vessel with a revolving turret fitted with two guns, she and the Confederate Navy's *Virginia* fought an inconclusive battle in the Chesapeake Bay on March 9, 1862.
[4]Ritchie Navigation is still in the business of making compasses, not in Brookline, but some 25 miles away in Pembroke, Massachusetts.

The Italian Navy, by the 1880s, had its own liquid compass designed by one of its officers, Captain G. B. Magnaghi. This compact and beautiful design contained a complete correcting system. Gone were the two iron spheres of the Kelvin binnacle, which often hindered the taking of bearings: in their place were two flat reels, looking like fishing reels, made of brass on which was wound iron wire. The Austro-Hungarian Navy also had its own liquid compass, complete with an ingenious correcting system, designed by Lieutenant Joseph Von Peichel.

Francis Crow, whether he knew it or not, had solved two of the disturbing problems inherent in a liquid compass. One is a problem known as "liquid swirl," which is caused when a vessel makes a quick alteration in course. The compass bowl, naturally, also makes the quick turn, and then friction between the bowl and the liquid drags the liquid around and sets up disturbing eddies. The other disturbing effect occurs when the vessel rolls. Liquid passes from the underside of the compass card to the top of the card, and vice versa. These two disturbances can be eliminated by having a compass card whose diameter is somewhat smaller than the diameter of the inside surface of the bowl. Crow's 1813 compass had this feature, but it was never adopted by other compass manufacturers. It was not until the first decade of the twentieth century that the benefits of the smaller card were recognized and became a standard part of the well-designed liquid compass.

A story is told that the use of the smaller card, like Fleming's discovery of penicillin, was an accident. Commander Louis Chetwynd, appointed Superintendent of the Admiralty Compass Department in 1904, had ordered a bowl and compass card for some experiments. Much to his annoyance he found that the compass card was the wrong size, being smaller than the size normally fitted. In extreme irritation he spun the bowl around and noticed that the card remained remarkably steady. Unlike Fleming, he realized the importance of his discovery. A year later, liquid compasses with reduced-diameter cards were being tested in sea trials and gun trials. The reports were unanimous in their approval of Chetwynd's liquid compass.

This compass, with its smaller diameter card, held all the design rules formulated by Creak for the proper functioning of a liquid compass: multiple needles; point of suspension in the same horizontal plane as the card and the pivots of the gimbals; the center of flotation below the point of suspension; the center of gravity of the card with its needles below the center of flotation.

Packed in the modern marina, where the masts of sailing yachts sprout like clusters of saplings, can be seen a later development of the liquid compass. Perched on binnacles and inset into decks and bulkheads are dome-topped compasses. The advantage of the dome-top compass is a reduction in swirl error and a compass card magnified by the curved shape of the top. Another benefit is the elimination of external gimbals. When vessel and binnacle heel, the compass card stays horizontal but is able to rotate into the domed-top without catching: a feat impossible for the card of a nongimballed flat-topped compass.

Today's marine magnetic compass has finally reached a state of perfection unknown to those early sailors of Alexander Neckham's day. Ironically, in a strange fashion, the marine magnetic compass has come full circle: from a magnetic needle floating in a bowl of water to a cluster of magnetic needles pivoting in a bowl of alcohol. But during its long evolution, it changed the trading patterns of the West, gave precision to sailing directions and charts, and—a small attribute never to be forgotten on the personal scale—gave comfort and reassurance to generations of sailors during those dark and never-ending hours of the middle watch between midnight and dawn.

This was the instrument that guided the Spanish 85-ton *Victoria* on the first circumnavigation of the world: a voyage accomplished with three simple instruments—astrolabe, quadrant, and compass.

From 1522, the year that the weather-beaten *Victoria* came to anchor off the quay at Seville, home from her historic circumnavigation, the whole world was magically transformed into an oyster for traders, merchants, shipowners, hydrographers, explorers, and mariners: an oyster ready and waiting to be opened, not with a sword, but by a compass needle.

EPILOGUE:
FROM NEEDLE TO
SPINNING TOP

For close to a thousand years the magnetized compass needle, no matter how it was suspended in its bowl to seek out the north magnetic pole, was the seaman's guide across the seas and oceans that make up two-thirds of our planet. Although at times an imperfect guide, given to wayward behavior from strange and mysterious influences, it was the sole instrument on which sailors placed their faith and their lives. Neglected and taken for granted it could, like a slighted god or goddess of the ancients, bring a fearful revenge in the shape of shipwreck and death.

But in the early years of the twentieth century the magnetic compass, now at its apotheosis as a magnificent navigational instrument, was unseated from its throne on the binnacle by a young usurper, the gyrocompass.

The gyrocompass is an electrically driven gyroscope, the

axis of which aligns itself along the meridian to point to true north. This distant relative of the child's spinning top also happens to be unaffected by both terrestrial magnetism and the unique magnetic signature of an individual ship. A marine compass that points to true north *and* is free from the sly influences of steel and iron, thus slaying at one stroke the compass errors and navigation mistakes brought about by variation and deviation, was bound to be an instrument that seamen would embrace with enthusiasm.

The gyrocompass started its life in 1901 with a proposal by the German engineer Dr. Hermann Anschütz-Kaempfe that a directional gyroscope could be used in a submarine to travel under the ice to the north pole. Magnetic compasses in this new and sinister type of warship presented a serious navigation problem, for a magnetic compass set within the cigar-shaped hull of a submarine is useless. The early submarines had their magnetic compass placed outside the pressure hull in a watertight binnacle, the image of the lubber line and compass card being projected to the steering position by an eye-straining telescopic system.

By 1908, Anschütz-Kaempfe had developed his idea of a directional gyroscope into the first gyrocompass, which was then promptly tested aboard the German battleship *Deutschland*. By 1911, Dr. Elmer Sperry, that most inventive of Americans with over four hundred inventions to his name, had produced his version of a gyrocompass. By the 1920s the world's navies were all fitting gyrocompasses (the Royal Navy opting for a Sperry model manufactured in England), for not only did they have this magic ability to point to true north but they could also run repeater compasses from the master compass hidden in the ship's bowels, keep a record of the course, and, most importantly, run an automatic helmsman.

In 1947, HMS *Vanguard*, sailing from England to Cape Town, suffered a complete power failure. All the whirling fly-wheels of her three gyrocompasses, deprived of the pulse that gave them life, toppled and stopped like a child's spinning top. The *Vanguard* had no back-up magnetic compasses due to an Admiralty decision taken in 1946. And so Britain's last battle-ship, the pride of the Navy, was reduced to steering by the stars—all the more embarrassing as King George VI happened to be aboard. Today the world's navies, both fighting and mer-cantile, carry magnetic compasses as a fail-safe measure when similar disasters strike.

The gyrocompass is a heavy, complicated, and expensive item of machinery. But for the weekend sailor in love with technology there is a lighter and cheaper compass: the fluxgate or inductor compass. Developed for aircraft, this subtle instru-ment seeks out the strength of the earth's magnetic field, rather like the way the sensory whiskers of a cat lead it to its stomach's gratification, and indicates not true north, as does the gyrocompass, but magnetic north. However, like the gyro-compass, the fluxgate compass requires an electric current—which takes us back to the fluxgate compass aboard the high-tech yacht described in the Prologue to this book.

In the spring of 1889 the Superintendent of the Admiralty Compass Department, Captain E. W. Creak, gave a talk on the marine magnetic compass to a group of naval officers. He ended his address with some sage advice to his listeners:

After all that has been said, true science teaches us a never-to-be-forgotten lesson. Observe the deviation when possi-ble, note it in a book, and learn its changes under every circumstance. . . . Then on some thick dirty night, the reward will come in the shape of a reasonable confidence that you can steer the required course without delay and detriment to the service upon which the ship may be ordered.

This advice holds as well as today as it did in 1889—even in a world of gyrocompasses and fluxgate compasses. For the marine magnetic compass, the sailor's guide, remains the essential instrument when all the others have failed.

APPENDIX: DEVIATION

The magnetic needle of the marine compass is held hostage to a host of hidden forces, deviation being, perhaps, the most curious and intractable. The word *deviation*, coined by Sir John Ross in 1820, defines the error caused by the attraction of a ship's ironwork. Since Sir John's day the deviation of the compass needle from pointing to the magnetic north has been analyzed and the causes placed in separate catagories.

Semicircular Deviation: The built-in magnetism of an iron or steel ship has three components: the fore-and-aft, the athwartship, and the vertical. When a ship is steering north or south, the fore-and-aft magnetism is in line with the compass needle and the needle has no deviation. When steering east or west the needle is deflected, making its maximum deviation. This deviation can be corrected with fore-and-aft magnets placed in the binnacle.

When a ship is steering east or west, the athwartship magnetism is in line with the compass needle and the needle has no deviation. When steering north or south the needle is deflected, making its maximum deviation. This deviation can be corrected by athwartship magnets placed in the binnacle.

Heeling Error: When a vessel heels, the compass card remains horizontal and the mass of steel or iron of the ship's hull and superstructure moves around it, thus changing the ship's magnetic field in relation to the compass needle. This, particularly when a ship is rolling heavily in a seaway, can send the compass card into wild oscillations. It was probably heeling error that set the *Ironside*'s steering compass (see Chapter 16) into its gyrations in heavy weather. Heeling error is compensated for by having an adjustable magnet hung below the compass.

Soft Iron and Hard Iron: The iron or steel in a ship, for the purposes of compass correction, can be considered as either magnetically *soft* or magnetically *hard*. *Soft iron* is easily magnetized but does not retain its magnetism. Align an iron poker in the earth's magnetic field so that it points North and South. The north end of the poker will become magnetized and North-seeking. Turn the poker East and West and it will lose its magnetism. Turn the poker again so that it aligns North and South, but with the ends reversed. The poker will become magnetized with the new end becoming North-seeking. The characteristic feature of *hard iron* is that it is difficult to magnetize, but once magnetized it retains its magnetism. Buy a magnet in a shop and you are buying *hard iron*.

Quadrantal Deviation: This deviation is caused by the magnetism induced into the horizontal "soft" iron of the ship by the horizontal component of the earth's magnetic field. An alteration of the ship's course will change this induced magnetism which, in turn, will increase or decrease the deviation. This error is compensated for by having two soft-iron spheres placed in brackets at the level of the compass card, either side of the binnacle. These spheres are adjustable and form the most conspicuous feature of the binnacle.

Sextantal Deviation and Octantal Deviation: These deviations, identified by Archibald Smith and Frederick Evans, are caused when a single, long compass needle is placed close to a compensating magnet or soft-iron corrector. When a ship is swung through 360° the compensating magnet will produce six points of maximum deviation (sextantal deviation). The induced magnetism in the soft-iron corrector will produce eight points of maximum deviation (octantal deviation). But, happily, these deviations are eliminated if multiple needles, as advocated by Smith and Scoresby, replace the single needle.

Flinders Bar: Matthew Flinders' great contribution to compass correction, a vertical bar of soft iron with its upper end at the level of the compass needle, corrects for the induced magnetism of the ship's vertical soft iron.

NOTES

Abbreviations in the Notes

ADM   Admiralty Records
BAAS   British Association for the Advancement of Science
BL   British Library
DNB   *Dictionary of National Biography*
HRNSW Historical Records New South Wales
PRO   Public Record Office, Kew
RS   Royal Society, London
          RSJB: Journal Book
          RSPT: Philosophical Transactions

PROLOGUE

p. 16   *"doesn't work yet"*: quoted in the *Sunday Times* (London), August
         29, 1999.
p. 17   *mast with a dagger:* Magnus, vol. II, pp. 495–96.

CHAPTER 1. DEAD RECKONING

p. 20   *rings on his fingers:* Larm, p. 44, and *DNB.*

p. 20   *"practicable and useful":* quoted in Howse, p. 52.

p. 21   *of HMS* Lennox *disagreed:* Larm, p. 43.

p. 21   *run up Channel:* May, "The Last Voyage of Sir Clowdisley Shovel," pp. 324–25.

p. 23   *"with red shells therein":* quoted in Taylor, *The Haven Finding Art,* p. 132.

p. 24   *"reckonings ordinarily made":* quoted in Waters, *The Art of Navigation,* p. 592.

p. 25   *in good working condition:* May, *History of Marine Navigation,* p. 76.

p. 26   *"than an old Groat":* quoted in May, "Naval Compasses in 1707," p. 406.

p. 26n   *until 1830:* May, *History of Marine Navigation,* p. 76.

p. 29   *"hard and fast aground":* quoted in Cowan, p. 59.

CHAPTER 2. NEEDLE AND STONE

p. 32   *underneath the plate:* Benjamin, pp. 87–88.

p. 32   *"a lamb from a wolf":* quoted in Taylor, *The Haven Finding Art,* p. 93.

p. 33   *play havoc with . . . the ship's navigation:* The unusual qualities of the lodestone can be found in Burton, *The Anatomy of Melancholy,* Part 2, p. 219; Hitchins and May, p. 20; Benjamin, pp. 219–20; May, "Garlic and the Magnetic Compass," pp. 231–34.

p. 33   *Alexander at the left:* Hitchins and May, p. 20.

p. 34   *"looks to the North":* quoted in Benjamin, pp. 128–29.

p. 34   *"vicissitudes of the atmosphere":* quoted in Benjamin, p. 129.

p. 34   *"it does not move":* quoted in Taylor, *The Haven Finding Art,* p. 94.

p. 35   *"travelling by sea":* quoted in Taylor, *The Haven Finding Art,* pp. 94–95.

p. 35   *"an art that never fails":* quoted in Taylor, *The Haven Finding Art,* pp. 95–96.

p. 36   *point to the south:* see Mitchell, chapter I, p. 110, and chapter II, p. 242; Needham, *Science and Civilization in China*, vol. 4, part 1, pp. 249–50.

p. 37   *its tail south:* Needham, *Science and Civilization in China*, p. 257.

p. 37   *"determine their whereabouts":* quoted in Needham, *Science and Civilization*, p. 279.

p. 37   *"belly of a shark":* quoted in Needham, *Science and Civilization*, p. 284.

p. 38   *"common in Chiangsu":* quoted in Needham, *Science and Civilization*, p. 289.

p. 39   *a king's treasury:* Taylor, *The Haven Finding Art*, p. 101.

CHAPTER 3. THE ROSE OF THE WINDS

p. 43   *laid-up alongside quays:* Casson, *The Ancient Mariners*, p. 39.

p. 43   *Cyprus coast about 300 B.C.:* Casson, *Ships and Seafaring*, p. 106; Gillmer, pp. 125–32.

p. 43   *400,000 tons of grain:* Casson, *Ships and Seafaring*, p. 123.

p. 44   *"now scared to death":* quoted in Hornell, p. 230.

p. 44   *"is 3,800 stadia with Boreas":* quoted in Taylor, *The Haven Finding Art*, p. 53.

p. 48   *off Sardinia's Cagliari Bay:* Crone, p. 35.

p. 48   *dated close to 1275:* Crone, p. 30.

p. 49   *such a loxodromic chart:* Thompson, *The Rose of the Winds*, p. 193.

p. 50   *"shattered by the compass":* Lane, p. 608.

p. 51   *by the Admiralty Hydrographer:* Fanning, p. xlii.

p. 53   *"his Memory on all Occasions":* Lever, p. 75.

CHAPTER 4. VARIATION AND DIP

p. 55n   *"my noble Intelligencer":* Deacon, pp. 1–4.

p. 56   *"wicked and damned spirits":* DNB.

p. 57   *"more than £20 given":* quoted in Praagh, p. 369.

p. 57   *compasses "of divers sorts":* quoted in Waters, *The Art of Navigation*, p. 530.

p. 59   *"from the north to the west"*: quoted in Taylor, *The Haven Finding Art*, p. 196.

p. 60   *"unlearned Mechanician"*: quoted in Taylor, *The Haven Finding Art*, p. 204.

p. 61   *"Stroken with choler"*: quoted in Taylor, "An Elizabethan Compass Maker," p. 40.

p. 63   the sailors' *"great peril"*: quoted in Waters, p. 156.

p. 65   *"observations which he has made"*: quoted in Thompson, *Gilbert of Colchester*, pp. 10–11.

CHAPTER 5. EDMOND HALLEY, POLYMATH

p. 69   a nearby diving bell: *RSJB* vol. 8, September 23, October 7, October 21, 1691.

p. 70   mistakes of previous charts: Cook, pp. 232–23.

p. 70   *"A woman might piss it out"*: quoted in Porter, *London: A Social History*, p. 105.

p. 71   *"he was removed to Oxford"*: quoted in Ronan, p. 6.

p. 72   *"Astronomical Observator and Assistant"*: quoted in Howse, p. 28.

p. 72n   rival the *"Weeping Willow"*: Falkus, p. 144.

p. 74   *"have fitting assistance"*: quoted in Cook, pp. 62–66.

CHAPTER 6. TO COMPASS THE GLOBE

p. 78   *"through the great South Sea"*: quoted in Thrower, *The Three Voyages of Edmond Halley*, p. 250.

p. 78   *"publick an undertaking"*: quoted in Thrower, *The Three Voyages*, p. 252.

p. 79   *"were a little less rustic"*: quoted in Massie, p. 177.

p. 80   people *"right nasty"*: Bowle, p. 220.

p. 81   and fishing gear: Thrower, *The Three Voyages*, p. 265.

p. 81   *"Cape of Good Hope"*: quoted in Thrower, *The Three Voyages*, p. 269.

p. 82   *"King's colours"*: quoted in Thrower, *The Three Voyages*, p. 279.

p. 83 *"displeasing and uneasy"*: quoted in Thrower, *The Three Voyages*, p. 282.

p. 83 *"abusive language and disrespect"*: quoted in Thrower, *The Three Voyages*, p. 287.

p. 83 *"under such circumstances"*: quoted in Thrower, *The Three Voyages*, p. 286.

p. 84 entirely on *"Divine Authority"*: quoted in Ronan, pp. 168–69.

p. 84 *"Submissive to their Masters"*: quoted in Thrower, *The Three Voyages*, p. 34.

p. 85 *"in case of extremity"*: quoted in Thrower, *The Three Voyages*, p. 297.

CHAPTER 7. HALLEYAN LINES

p. 88 *"to Saile like the Wind"*: quoted in Thrower, *The Three Voyages*, p. 127.

p. 88 *"my poor boy Manley White"*: quoted in Thrower, *The Three Voyages*, p. 130.

p. 88 *"were backward the last"*: quoted in Thrower, *The Three Voyages*, p. 305.

p. 89 *"Scarce seen this fortnight"*: quoted in Thrower, *The Three Voyages*, p. 166.

p. 90 *"she wrighted again"*: quoted in Thrower, *The Three Voyages*, p. 174.

p. 91 *"extraordinary care of my Doctor"*: quoted in Thrower, *The Three Voyages*, p. 308.

p. 92 *"laid down with Marks"*: RSJB vol.8, October 30, 1700.

p. 94 *"without Variation"*: This is taken from Halley's Chart.

p. 94 *"alter the whole System"*: quoted in Thrower, *The Three Voyages*, p. 366.

p. 95 *"of their Expectations"*: An Advertisement is reproduced as a plate in Ronan and is quoted in Thrower, pp. 64–65.

p. 96 *"brandy like a sea-captain"*: this is a favorite quotation when it comes to Halley. It is quoted in Thrower, *The Three Voyages*, p. 74; Ronan, p. 185; and Cook, p. 321.

p. 98   *"towards their Improvement":* quoted in Taylor, *The Haven Finding Art*, p. 240.

CHAPTER 8. DR. GOWIN KNIGHT AND HIS MAGNETIC MACHINE

p. 99   *most of the sails:* Waddell, pp. 652–53.

p. 100   *in the Isle of Wight:* Waddell, p. 653.

p. 101   *used in its construction:* Knight, "Of the Mariner's Compass," p. 654.

p. 101   *a weight of 700 grains:* Harris, p. 6.

p. 102   *"more accurat & exact":* This card is reproduced in Fara, p. 32.

p. 102n   *"of my whole life":* Gibbon, p. 36.

p. 104   *"the most distant climes":* RSJB, vol. 19, November 30, 1747, pp. 359–66.

p. 104   *Thursday evening dining club:* Allibone, p. 32.

p. 105   *bar or compass needle placed between them:* Fothergill, pp. 117–20.

p. 105n   *"make thee happy":* Fox, pp. 220–21.

p. 107   *"exhilarating force":* Porter, *Health for Sale*, p. 161.

p. 107   *"master of his doctrine":* quoted in Fara, p. 44.

CHAPTER 9. KNIGHT'S COMPASS

p. 110   *"observed throughout the whole":* Knight, "Of the Mariner's Compass," p. 655.

p. 112   *"We must have at it again":* DNB.

p. 114   *"pretend to make extempore":* Knight, "Description of a Mariner's Compass," p. 67.

p. 114   *a very angry father:* Smiles, p. 107.

p. 114   *other than the law:* Smiles, p. 110.

p. 115n   *"are committed":* George, p. 351.

p. 116   *new compass as a starting point:* Smeaton, "On Some Improvements of the Mariner's Compass," pp. 67–70.

p. 116   *manner of taking a bearing:* Hitchins and May, p. 33.

CHAPTER 10. THE SHOCKS OF
TEMPESTUOUS SEAS

p. 120   *over the Spanish:* Williams, p. 247.

p. 120   *"may be greatly promoted":* Anson, 1911.

p. 121   *reported their findings:* PRO, ADM 106/2077, Admiralty Secretary to Navy Board, February 28, 1751; PRO, ADM 106/2185, Navy Board to Admiralty Secretary, March 27, 1751; PRO, ADM 3/62, Admiralty Minutes Book.

p. 121   *"made on the compass":* PRO, ADM 2/73. Letters from Admiralty Board to captains, April 4, 1751.

p. 121   *"bring them back to England":* PRO, ADM 2/73, Letter from Admiralty Board to Wyatt, April 7, 1751.

p. 122   *trolling for fish:* Smeaton, "An Account of Some Experiments . . . ," pp. 532–48.

p. 122   *"a considerable merchant-ship":* Smeaton, "An Account of Some Experiments. . . ."

p. 122   *"beer-drawer for the men":* DNB.

p. 123   *"strict devotion":* DNB.

p. 123   *"than the old compass":* PRO, ADM 51/361, Logbook of the *Fortune*, September 30, 1751.

p. 123   *"except in stormy weather":* PRO, ADM 106/2186, Letter from Navy Board to Admiralty Board, April 29, 1752.

p. 123   *"impossible to steer by":* Spinney, p. 107.

p. 124   *compass and magnet bar:* PRO, ADM 3/62, Admiralty Board Minutes, June 24 and July 29, 1752.

p. 126   *a lozenge-shaped needle:* A list of eighteenth-century compass prices is given in Fara, p. 83.

p. 126   *"rightly constructed":* quoted in Fara, p. 82.

p. 126   *"that important instrument":* Hutchinson, p. 104.

p. 126   *"degree of Perfection":* quoted in Taylor, *The Haven Finding Art*, pp. 240–41.

p. 126   *"celebrated Dr. Gowin Knight":* Robertson, *The Elements of Navigation*, p. 232.

p. 127 *"the Motion of the Ship"*: BL, Patent No. 850, 1766.

p. 127 *"are Intended for"*: G. Robertson, p. 5.

p. 127 *"Roals or pitches much"*: G. Robertson, p. 7.

p. 128 *"neaver Stands Steedy"*: G. Robertson, p. 29.

p. 128 *"they are generally useless"*: PRO, ADM 106/1163, Letter from Cook to Navy Board, March 8, 1768.

p. 129 *"Transit of Venus"*: RS Council Minutes, May 5, 1768.

p. 130 *"to be supplyed with it"*: PRO, ADM 1/1609, Letter from Cook to Admiralty Board, July 25, 1768.

p. 131 *"Compass in a troubled Sea"*: PRO, ADM 1/1609, Letter from Cook to Admiralty Board, July 12, 1771.

p. 132 *"women with child"*: quoted in Gunther, p. 158.

p. 132 *in a single group:* Edwards, p. 324.

p. 132 *"on the way to it"*: Toynbee and Whibley, vol. 2, pp. 631–32.

p. 133 *"circumstances attending it"*: Stackpole, p. 248.

p. 133 *"and subject to vibrate"*: PRO, ADM 106/3401, Letter from Deptford Yard to Navy Board, August 12, 1768.

CHAPTER 11. ANY OLD IRON, ANY OLD IRON

p. 136 *off St. Alban's Head:* Ingleton, p. 116.

p. 136 *from the Board of Longitude:* May, "Longitude by Variation," pp. 339–41.

p. 137 *Royal Gardens at Kew:* Ingleton, p. 16.

p. 138 *proficient in its use:* May, "Longitude by Variation," p. 341.

p. 139 *"will agree exactly"*: quoted in Fanning, p. xxiii.

p. 139 *"anchors and other iron"*: quoted in Fanning, p. xxii.

p. 140 *"errors than this"*: quoted in Beaglehole, *The Journals of Captain James Cook*, vol. III, part 1, p. 23.

p. 140 *"materially from one another"*: quoted in Hewson, pp. 57–58.

p. 141 *as an able seaman:* Kippis, p. 3; Beaglehole, *The Life of Captain James Cook*, p. 15.

p. 142 *"tipping a heiva"*: quoted in Beaglehole, *The Life of Captain James Cook*, p. 711.

p. 142 *"reading Robinson Crusoe"*: *Naval Chronicle*, vol. 32 (1814), p. 178.

p. 142n *"paroxysms of passion"*: quoted in Beaglehole, *The Life of Captain James Cook*, p. 711.

p. 143 *Walker azimuth compass*: HRNSW vol. II, p. 255. Letter from Hunter to Stephens, August 9, 1794.

p. 143 *"in the said list"*: HRNSW vol. II, p. 255. Letter from Stephens to Hunter, September 5, 1794.

p. 143 *"ship's boats upon"*: HRNSW vol. II, p. 257. Letter from Stephens to Hunter, October 10, 1794.

p. 144 *schooner's course was altered*: Flinders, *A Voyage to Terra Australis*, vol. I, p. cxxvi.

p. 146 *sail on foreign duties*: PRO, ADM 2/293 and 2/294, Letters from Admiralty Board to Navy Board, November 21 and December 10, 1800.

p. 147 *their commissioned commander*: PRO, ADM 55/75, Flinders, "Journal" vol. I, p. 1.

CHAPTER 12. THE BOOK OF BEARINGS

p. 150 *objective was the southwest coast of New Holland*: Brown, p. 28; Ingleton, pp. 65–66.

p. 150 *approval of Napoleon Bonaparte*: Brown, p. 28.

p. 150 *applying for the passport*: Mack, pp. 75–77.

p. 151 *"entirely to your decision"*: PRO, ADM 2/295, Letter from Nepean to Banks, April 28, 1801.

p. 152 *used a Flinders chart*: Moorehead, p. 162.

p. 152 *"left near the binnacle"*: Flinders, *Voyage*, vol. I, p. 129.

p. 154 *"the ship into harbour"*: Flinders, *Voyage*, vol. I, p. 226.

p. 155 *for the French officers*: Brown, p. 235.

p. 155 *their two countries*: HRNSW vol. IV, p. 949, Letter from Philip Gidley King to Baudin, June 24, 1802.

p. 158 *"the generality of seamen"*: HRNSW vol. III, p. 724, Letter from Philip Gidley King to John King, Under Secretary of the Colonial Department, October 8, 1799.

p. 158 *"to render her assistance"*: Flinders, *Voyage*, vol. II, p. 96.

p. 159 *"without much risk"*: Flinders, *Voyage*, vol. II, p. 145.

CHAPTER 13. THE FLINDERS BAR

p. 161 *rotten and unseaworthy:* Flinders, *Voyage,* vol. II, p. 274.

p. 164 *"moments of my life":* Flinders, *Voyage,* vol. II, p. 327.

p. 165 *"future expected voyage":* Flinders, *Voyage,* vol. II, pp. 351–53;
PRO, ADM 55/75, Flinders "Journal" vol. III.

p. 165 *detested the English:* Brown, p. 371.

p. 166 *"of the Ship's Head":* RSPT vol. 95 (1805).

p. 166 *in marine meteorology:* RSPT vol. 96 (1806).

p. 167 *gods of their bellies:* Ingleton, p. 261.

p. 170 *"reach of General De Caen":* Flinders, *Voyage,* vol. II, p. 485.

p. 171 *Portsmouth, and Plymouth:* PRO, ADM 1/1809, Letter from
Flinders to Croker, 13 April 1812.

p. 171 *"the risk of being neglected":* quoted in Ingleton, p. 410.

p. 172 *"particular marline spikes":* PRO, ADM 1/1809/130A.

p. 172 *"with the compass card":* PRO, ADM 1/1809/130A.

p. 172 *old man of seventy:* Ingleton, p. 420.

CHAPTER 14. SOFT IRON, HARD IRON

p. 176 *"needle of the compass":* quoted in Stamp, p. 19.

p. 176 *"with divine writings":* quoted in Stamp, p. 14.

p. 177 *his crew's tobacco pipes:* Scoresby, *An Account of the Arctic Regions*
. . . , vol. I, pp. 282–83.

p. 177 *"must necessarily perish":* Scoresby, *Arctic Regions,* vol. I, p. 546.

p. 177 *"confining the toes":* Scoresby, *Arctic Regions,* vol. II, p. 269.

p. 179 *"for many years past":* Scoresby-Jackson, p. 123–24.

p. 179 *"subordinate capacity":* Scoresby-Jackson, p. 123.

p. 180 *"to that intended":* Scoresby, *Arctic Regions,* vol. II, p. 552.

p. 182 *a certain Dr. Thomas Young:* Young, vol. II, pp. 102–10.

p. 182 *"physiological optics":* DNB.

p. 183 *sailing parallel . . . was impossible:* Johnson, p.153.

p. 183 *cold, heat, and wind:* Scoresby, *Arctic Regions,* vol. II, p. 545.

p. 185 *"beginning of the 18th century":* PRO, ADM 106/2531.

p. 185 *"as it is at present":* PRO, ADM 106/1465.

CHAPTER 15. "AN EVIL SO PREGNANT WITH MISCHIEF"

p. 187   *wear cork life jackets:* Fairbairn, pp. 138–39.

p. 188   *18 inches of water:* Fairbairn, p. 136.

p. 189   *"without further mishap":* Fairbairn, p. 141.

p. 190   *close to the binnacle:* For details on these shipwrecks, see Johnson, pp. 9, 157; Harris, p. 165; Fanning, p. xxxi; Gilly, pp. 59–69, 302–15; *The Times* (London), November 23, 1842.

p. 191   *"attraction of the vessel":* Laird and Oldfield, vol. II, p. 153.

p. 193   *"the magnetic meridian":* RSPT vol. 126 (1836).

p. 193   *"correcting for the deflection":* RSPT vol. 126 (1836).

p. 193n   *"compass on deck":* Scoresby, *Arctic Regions*, vol. II, p. 549.

p. 195   *"has not been the result":* quoted in Fanning, p. 1.

p. 195   *"an evil so pregnant with mischief":* quoted in Fanning, p. 3.

p. 197   *"in chronometer business":* Airy, *Autobiography*, p. 124.

p. 197   *"to master their contents":* Airy, *Autobiography*, pp. 2–3.

p. 197   *buildings at Greenwich:* Airy, *Autobiography*, p. 133.

CHAPTER 16. DEVIATION, THE HYDRA-HEADED MONSTER

p. 200   *three other compasses:* Fanning, p. xxxvii; Airy, *Autobiography*, p. 135.

p. 200   *"sensibly correct":* Airy, *Autobiography*, p. 135.

p. 201   *astigmatic St. George:* Airy, *Autobiography*, p. 1.

p. 201   *iron-hulled sailing ship:* Airy, *Autobiography*, p. 135; Cotter, "George Biddle Airy," pp. 266–67.

p. 201   *"have no interest":* quoted in Cotter, "George Biddle Airy," p. 267.

p. 201   *"every part of the earth":* quoted in Cotter, "George Biddle Airy," p. 267.

p. 203   *"ocean-going vessels":* *Liverpool Albion*, February 18, 1839.

p. 204   *when under sail:* Creuze, pp. 90–91.

p. 204   *"private armed steamer":* Bernard, p. 2.

p. 205 *"uninstructed common sense":* Creuze, p. 100.

p. 206 "of the Nemesis": Airy, "On the Correction of the Compass,"
pp. 239–43.

p. 207 *unfortunate Mr. Cocking: The Times* (London), July 25, 1837.

p. 207 *June 29, 1840:* Fanning, p. 11.

p. 208n *in use in 1944:* Fanning, p. 6.

p. 209 *he made his own:* RS Proceedings vol. 22 (1873–74), p. i.

p. 210 *spare compasses and cards:* Admiralty Memorandum of April 11,
1843, and Admiralty Circular of November 20, 1845, quoted in
Fanning, pp. 427–30.

CHAPTER 17. THE "INEXTRICABLE ENTANGLED WEB"

p. 211 *"to the British Museum":* Hargraves, p. 116.

p. 212 *"highest rate of speed":* quoted in Barnaby, p. 57.

p. 213 *lasted two days:* BAAS Report No. 24, pp. 29–50.

p. 213 *and Physical Science:* Scoresby-Jackson, p. 259.

p. 214 *"high admirer of female beauty":* quoted in Scoresby-Jackson,
p. 95.

p. 214 *"he wished, I wished":* quoted in Stamp, p. 214.

p. 214 *"adopt the plan":* quoted in Stamp, p. 133.

p. 215 *that of Dr. Gowin Knight:* Stamp, pp. 137–38.

p. 215 *"Compass in Iron Ships": The Athenaeum* No. 1406, October 7,
1854.

p. 216 *"blinking the question":* quoted in Scoresby-Jackson, p. 352.

p. 217 *"alarmist doctrines of Dr. Scoresby":* Airy, *The Athenaeum* No. 1423,
February 3, 1855, pp. 145–48.

p. 218 *"entangled web":* quoted in Fanning, p. 65.

p. 219 *"most pernicious":* quoted in Smith and Wise, p. 769.

p. 220 *construction and magnetism:* Scoresby, *Journal of a Voyage to Australia*
. . . , p. 19.

p. 221 *"polarity instead of southern":* Scoresby, *Journal of a Voyage*, p. 158.

p. 221 *for iron-hulled ships:* Scoresby, *Journal of a Voyage*, p. 184.

p. 222 *"reference at all times":* quoted in Scoresby-Jackson, p. 384.

CHAPTER 18. GRAY'S BINNACLE

p. 226   *"of course be extinguished"*: quoted in Wright, p. 188.

p. 226   *"principles of its construction"*: Airy, *Autobiography*, p. 207.

p. 226   *in the Exhibition*: Fanning, p. 34.

p. 228   *"of vessels of war"*: quoted in Watts, *The Royal Navy*, p. 19.

p. 228   *"stubbornly resists change"*: Fisher, *Records*, p. 177.

p. 228   *"that approaches them"*: quoted in Rolt, *Isambard Kingdom Brunel*, p. 242.

p. 229   *lay with compass deviation*: Rolt, *Isambard Kingdom Brunel*, p. 228.

p. 230   *"so tortured in its action"*: quoted in Fanning, p. 64.

p. 233   *"latest requirements of science"*: *The Times* (London), January 26, 1866.

CHAPTER 19. THOMSON'S COMPASS AND BINNACLE

p. 236   *"science of this country"*: quoted in Thompson, *Life of Lord Kelvin*, vol. II, p. 632.

p. 236   *"as I now do"*: quoted in Smith and Wise, p. 755.

p. 236n   *"I daresay her majesty"*: quoted in Smith and Wise, p. 762.

p. 238   *"these five years"*: quoted in Thompson, *Life*, vol. II, p. 698.

p. 238   *"material can give"*: quoted in Thompson, *Life*, vol. II, p. 586.

p. 239   *"of an inch long"*: quoted in Thompson, *Life*, vol. II, p. 704.

p. 239   *"movable compass card"*: quoted in Thompson, *Life*, vol. II, p. 703.

p. 239   *"suggestions from without"*: quoted in Thompson, *Life*, vol. II, p. 705.

p. 239   *was the waved message*: Thompson, *Life*, vol. II, pp. 638–39.

p. 240   *"to write legibly"*: quoted in Thompson, *Life*, vol. II, p. 669.

p. 242   *adjustable soft-iron spheres*: Thomson, "On Compass Adjustment in Iron Ships," pp. 91–119.

p. 243   *encased in a brass tube*: Thomson, "Recent Improvements in the Compass," pp. 404–14.

p. 243   *"by one colossal railway"*: quoted in Judd, p. 93.

CHAPTER 20. THE SELLING OF A COMPASS

p. 248   *"was going on shore"*: quoted in Thompson, *Life of Lord Kelvin*, vol. II, p. 592.

p. 249   *"off my waterproof"*: quoted in Thompson, *Life*, vol. II, pp. 612–16.

p. 250   *tonnage built in Britain*: Pollard and Robertson, *The British Shipbuilding Industry*, pp. 61–62.

p. 250   *ordered from abroad*: Bauer, pp. 289–90.

p. 250   *rigors of the sea*: Smith and Wise, p. 782.

p. 251   *"known compass possesses"*: Lecky, *"Wrinkles" in Practical Navigation*, p. 11.

p. 252   *introduced into the Royal Navy*: This letter is given in Thompson, *Life*, pp. 711–12.

p. 253   *"none was asked for"*: quoted in Fanning, p. 104.

p. 253   *"position of the ship"*: Thomson, "On Compass Adjustment in Iron Ships," pp. 92–95.

p. 254   *"John Arbuthnot Fisher"*: Bacon, vol. I, p. 6.

p. 254   *damn the breakages*: Morris, p. 113.

p. 256   *to merchant ship owners*: *The Times* (London), November 19, 1879; Smith and Wise, p. 792.

p. 256   *by a liquid compass*: Fanning, p. 115.

p. 256   *"anything about this"*: quoted in Smith and Wise, p. 795; Fisher, pp. 62–63.

p. 256   *sent to Glasgow*: Fanning, p. 119.

p. 257   *"Navy is concerned"*: Bacon, vol. I, p. v.

p. 257   *"of my existence"*: quoted in Fanning, p. 154.

p. 257   *"his compass installed"*: RS Proceedings, vol. 97 (1920), p. xiii.

p. 258   *"they came from us"*: quoted in Smith and Wise, p. 797.

p. 258   *to compass matters*: Fanning, p. 75.

p. 259   *were in grave danger*: Armstrong, p. 166.

CHAPTER 21. A QUESTION OF LIQUIDITY

p. 261   *added another 825*: Armstrong, p. 189.

p. 263   *world's fastest ships*: Watts, p. 58. Armstrong, pp. 228–61.

p. 263   *"quite unserviceable":* quoted in Fanning, p. 114.

p. 263n *"destroy the French boats":* quoted in Bacon, vol. I, p. 110.

p. 264   *"the same conditions":* quoted in Fanning, p. 114.

p. 264   *"faults in their inventions":* Creak, p. 965.

p. 264   *"by Sir William Thomson":* Armstrong, p. 208.

p. 265   *"hither after all":* quoted in Williams, p. 43.

p. 266   *in a glass tube:* Ingenhousz, *RSPT* vol. 69 (1779), pp. 537–46.

p. 268   *"of the smallest size":* BL Patent No. 3644 (1813).

p. 268   *of the liquid compass:* Greater detail of the seagoing and gunfire tests is given in Johnson, pp. 77–82.

p. 269   *"naval or mercantile":* quoted in Creak, p. 965.

p. 270   *wound iron wire:* article in *Journal of the Royal United Service Institution,* vol. 39 (1895), pp. 211–12.

p. 271   *Chetwynd's liquid compass:* Fanning, p. 169.

EPILOGUE

p. 276   *"the ship may be ordered":* Creak, p. 966.

BIBLIOGRAPHY

Abbreviations in the Bibliography:

BAASR *British Association for the Advancement of Science Report*
JIN   *Journal of the Institute of Navigation*
JRUSI *Journal of the Royal United Services Institute*
NMM   *National Maritime Museum, Greenwich*
MM   *Mariner's Mirror*
RSPT   *Royal Society Philosophical Transactions*

Aczel, Amir D. *The Riddle of the Compass.* New York, 2001.
Airy, G. B. "On the Correction of the Compass in Iron-built Ships."
    United Services Journal (1840).
————. *Autobiography of Sir George Biddell Airy.* Edited by Wilfrid Airy.
    Cambridge, England, 1896.
Allibone, T. E. *The Royal Society and Its Dining Clubs.* Oxford, 1976.
Anson, George. *A Voyage Round the World.* London, 1911.
Armstrong, G. E. *Torpedoes and Torpedo Vessels.* London, 1896.

Ashburner, W. *The Rhodian Sea-Law*. Oxford, 1909.

Bacon, R. H. *The Life of Lord Fisher of Kilverstone*. 2 vols. London, 1929.

Bain, William. *An Essay on the Variation of the Compass*. Edinburgh, 1817.

Baker, S. J. *My Own Destroyer: A Biography of Capt. Matthew Flinders RN.* Sydney, 1962.

Barlow, William. *The Navigators Supply*. London, 1597.

——. *Magneticall Advertisements*. London, 1616.

——. *A Briefe Discovery*. London, 1618.

Barnaby, K. C. *Some Ship Disasters and Their Causes*. London, 1968.

Bauer, K. J. *A Maritime History of the United States*. Columbia, South Carolina, 1988.

Bayne-Powell, R. *Eighteenth-Century London Life*. London, 1937.

Beaglehole, J. C. *The Life of Captain James Cook*. London, 1974.

—— (editor). *The Journals of Captain James Cook*, volume 3, part 1. Cambridge, England, 1967 (Hakluyt Society Extra Series 34a).

Beazley, C. R. *The Dawn of Modern Geography*. 3 vols. London, 1897.

Benjamin, P. *The Intellectual Rise in Electricity*. London, 1895.

Bernard, W. D. *Narrative of the Voyages and Services of the Nemesis from 1840 to 1843*. London, 1844.

Bevan, Bryan. *Charles the Second's French Mistress*. London, 1972.

Blackbarrow, Peter. *The Longitude Not Found*. London, 1678.

Bond, Henry. *The Longitude Found*. London, 1676.

Borough, William. *A Discourse of the Variation*. London, 1581.

Bourne, William. *A Regiment for the Sea and Other Writings on Navigation*. Edited by E. G. R. Taylor. London, 1963 (Hakluyt Society Series 2/121).

Bowle, John. *John Evelyn and His World*. London, 1981.

Brown, A. J. *Ill-Starred Captains: Flinders and Baudin*. London, 2001.

Burchett, J. A. *A Complete History of the Most Remarkable Transactions at Sea*. London, 1720.

Burford, E. J. *Royal St James's*. London, 2001.

Burton, Robert. *The Anatomy of Melancholy*. London, 1932.

Canton, J. "A Method of making Artificial Magnets without the use of Natural Ones." *RSPT*, vol. 47 (1751).

Casson, Lionel. *The Ancient Mariners*. London, 1959.

————. *Ships and Seafaring*. London, 1994.

Chaucer, Geoffrey. *The Complete Works of Geoffrey Chaucer*. Edited by Walter W. Skeat. London, 1946. Contains "Treatise on the Astrolabe."

Cook, Alan. *Edmond Halley: Charting the Heavens and the Seas*. Oxford, 1998.

Cooke, C. W. *William Gilbert of Colchester*. London, 1890.

Cotter, C. H. "George Biddell Airy and his Mechanical Correction of the Magnetic Compass." *Annals of Science*, vol. 33 (1976).

————. "The Early History of Ship Magnetism: The Airy-Scoresby Controversy." *Annals of Science*, vol. 34 (1977).

Cowan, Edward. *Oil and Water: The Torrey Canyon Disaster*. London, 1969.

Creak, E. W. "On the Mariner's Compass in Modern Vessels of War." *JRUSI*, vol. 33 (1889–90), pp. 949–75.

Creuze, A. F. B. "On the *Nemesis* Private Armed Steamer, and on the Comparative Efficiency of Iron-Built and Timber-Built Ships." *United Services Journal* (1840), pp. 90–100.

Crone, G. R. *Maps and Their Makers*. London, 1962.

Deacon, Richard. *John Dee*. London, 1968.

De Beer, Gavin. *The Sciences Were Never at War*. London, 1960.

Dee, John. *The Diaries of John Dee*. Edited by Edward Fenton. Charlebury, England, 1998.

Edwards, E. *Lives of the Founders of the British Museum*. 2 vols. London, 1870.

Evans, F. J. "Notes on the Magnetism of Ships." *JRUSI*, vol. 3 (1859–60), pp. 91–110.

————. "On the Magnetism of Iron and Iron-Clad Ships." *JRUSI*, vol. 9 (1865–66), pp. 277–98.

Evelyn, John. *Diary of John Evelyn*. 2 vols. London, 1966.

Fairbairn, W., and W. Pole. *The Life of Sir William Fairbairn*. London, 1877.

Falconer, William. *An Universal Dictionary of the Marine*. London, 1769.

Falkus, Christopher. *The Life and Times of Charles II*. London, 1992.

Fanning, A. E. *Steady As She Goes*. London, 1986.

Fara, Patricia. *Sympathetic Attractions*. Princeton, 1996.

Fisher, J. A. *Records*. London, 1919.

Flinders, Matthew. "Concerning the Difference in the Magnetic Needle on board the Investigator, arising from an Alteration in the Direction of the Ship's Head." *RSPT*, vol. 95 (1805).

————. "Magnetism of Ships." *Naval Chronicle*, vol. 28 (1812), pp. 318–24.

————. *A Voyage to Terra Australis*. 2 vols. London, 1814.

Forbes, E. G. "The Birth of Navigational Science." NMM, Monograph no. 10, 1974.

Fothergill, J. "An Account of the Magnetical Machine contrived by the late Dr. Gowin Knight." *RSPT*, vol. 66 (1776).

Fox, R. Hingston. *Dr. John Fothergill and His Friends*. London, 1919.

French, P. *John Dee: The World of an Elizabethan Magus*. London, 1972.

Friendly, Alfred. *Beaufort of the Admiralty: The Life of Sir Francis Beaufort 1774–1857*. New York, 1977.

George, W. D. *London Life in the Eighteenth Century*. London, 1930.

Gibbon, Edward. *Autobiography of Edward Gibbon*. London, 1972.

Gilbert, W. *The Loadstone*. Translated from the Latin by P. F. Mottelay. London, 1893.

Gillmer, Thomas. *A History of Working Watercraft of the Western World*. Camden, Maine, 1994.

Gilly, William. *Shipwrecks of the Royal Navy, 1793–1849*. London, 1850.

Gouk, Penelope. *The Ivory Sundials of Nuremberg*. Cambridge, England, 1988.

Gunther, A. E. *The Founders of Science at the British Museum 1753–1900*. Halesworth, 1980.

Halley, Edmond. *The Three Voyages of Edmond Halley in the Paramore 1698–1701*. Edited by N. J. W. Thrower. London, 1981 (Hakluyt Society Series 2/156).

————. "A Theory of the Variation of the Magnetical Compass." *RSPT*, vol. 13 (1683).

————. "An Account of the Cause of the Change of the Variation of the Magnetick Needle." *RSPT*, vol. 17 (1692).

————. "An Advertisement Necessary for all Navigators bound up the Channel of England." *RSPT*, vol. 22 (1700).

————. "Observations of Latitude and Variation, taken on board the Hartford." *RSPT*, vol. 37 (1732).

Hargraves, E. H. *Australia and Its Gold Fields: A Historical Sketch of the Progress of the Australian Colonies, from the Earliest Times, to the Present Day.* London, 1855.

Harris, W. Snow. *Rudimentary Magnetism.* 3 Parts. London, 1850–52.

Harrison, Edward. *Idea Longitudinis, being a Brief Definition of the best known Axioms for finding the Longitude.* London, 1696.

Hewson, J. B. *A History of the Practice of Navigation.* Glasgow, 1951.

Hibbert, Christopher. *The English: A Social History 1066–1945.* London, 1994.

Hill, H. O., and E. W. Paget-Tomlinson. *Instruments of Navigation.* London, 1958.

Hitchins, H. L., and W. E. May. *From Lodestone to Gyro-Compass.* London, 1955.

Hornell, James. *Water Transport: Origins and Early Evolution.* Newton Abbot, England, 1970.

Howse, Derek. *Greenwich Time and the Discovery of the Longitude.* Oxford, 1980.

Hutchinson, William. *A Treatise on Practical Seamanship.* London, 1777.

Ingleton, Geoffrey. *Matthew Flinders, Navigator and Chartmaker.* Guildford, England, 1986.

Inwood, Stephen. *A History of London.* London, 2000.

Johnson, E. J. *Practical Illustrations of the Necessity for Ascertaining the Deviations of the Compass.* London, 1852.

Judd, Denis. *Empire.* London, 1996.

Kippis, A. *A Narrative of the Voyages Round the World Performed by Captain James Cook.* London, 1893.

Knight, Gowin. "Description of a Mariner's Compass contrived by Gowin Knight, M.B., F.R.S." *RSPT*, vol. 46 (1749–50).

————. "Of the Mariner's Compass that was struck with Lightning, as related in the foregoing Paper; with some further Particulars relating to that Accident." *RSPT*, vol. 46 (1749–50).

Laird, Macgregor, and R. A. K. Oldfield. *Narrative of an Expedition into the Interior of Africa.* 2 vols. London, 1837.

Lamb, H. H. *Historic Storms of the North Sea, British Isles and North-West Europe.* Cambridge, 1991.

Lane, F. L. "The Economic Meaning of the Invention of the Compass." *American Historical Review*, vol. 68, no. 3 (1963).

Larm, Richard. *Cornish Shipwrecks: The Isles of Scilly*. Newton Abbot, 1969.

Lecky, S. T. S. *"Wrinkles" in Practical Navigation*. London, 1881.

Lever, Darcy. *The Young Sea Officer's Sheet Anchor*. London, 1808.

Liverpool Compass Committee. *Third Report of the Liverpool Compass Committee to the Board of Trade*. London, 1857–1860.

Mack, J. D. *Matthew Flinders 1774–1814*. Melbourne, 1966.

Magnus, Olaus. *A Description of the Northern Peoples 1555*. 3 vols. London, 1996 (Hakluyt Society Series 2/182).

Marcus, G. J. *The Navigation of the Norsemen. MM*, vol. 39 (1953).

————. "The Mariner's Compass: Its Influence upon Navigation in the Later Middle Ages." *History*, vol. 41 (1956).

————. "Dead Reckoning and the Ocean Voyages of the Past." *MM*, vol. 44 (1958).

————. *The Conquest of the North Atlantic*. Woodbridge, England, 1998.

Massie, Robert K. *Peter the Great*. London, 1995.

May, W. E. "The Birth of the Compass." *JIN*, vol. 2 (1949).

————. "The History of the Magnetic Compass." *MM*, vol. 38 (1952).

————. "Naval Compasses in 1707." *JIN*, vol. 6 (1953).

————. "The Binnacle." *MM*, vol. 40 (1954).

————. "Longitude by Variation." *MM*, vol. 45 (1959).

————. "The Last Voyage of Sir Clowdisley Shovell." *JIN*, vol. 13 (1960).

————. *A History of Marine Navigation*. London, 1971.

————. "Garlic and the Magnetic Compass." *MM*, vol. 65 (1979).

Mitchell, A. C. "Chapters on the History of Terrestrial Magnetism." *Terrestrial Magnetism and Atmospheric Electricity*. I. "On the directive property of a magnet in the earth's field and the origin of the nautical compass" vol. 37 (1932). II. "The discovery of the magnetic declination" vol. 42 (1937). III. "The discovery of the magnetic inclination" vol. 44 (1939).

Moorehead, Alan. *The Fatal Impact*. New York, 1966.

Morris, J. *Fisher's Face*. London, 1996.

Nansen, Fridtjof. *In Northern Mists.* 2 vols. London, 1911.

Needham, Joseph. *Chinese Science.* London, 1946.

————. *Science and Civilisation in China.* 12 vols. London, 1954–84.

Norman, Robert. *The Newe Attractive.* London, 1581.

————. *The Safeguard of Sailors.* London, 1584.

Parry, J. H. *The Age of Reconnaissance.* London, 1963.

Pollard, S., and P. Robertson. *The British Shipbuilding Industry 1870–1914.* London, 1979.

Porter, Roy. *Health for Sale: Quackery in England 1660–1850.* Manchester, 1989.

————. *London: A Social History.* London, 1996.

Praagh, G. Van. "John Dee (1527–1608)." *Discovery,* vol. 14 (1953).

Quinn, Paul. "The Early Development of Magnetic Compass Correction." *MM,* vol. 87 (2001).

Ritchie, G. S. *The Admiralty Chart.* Durham, 1995.

Robertson, George. *The Discovery of Tahiti: A Journal of the Second Voyage of HMS Dolphin round the World.* Edited by Hugh Carrington, London. 1948 (Haklyut Society Series 2/98).

Robertson, John. *The Elements of Navigation.* 2 vols. Sixth edition. London, 1796.

Rolt, L. T. C. *Isambard Kingdom Brunel.* London, 1961.

Ronan, C. A. *Edmond Halley: Genius in Eclipse.* London, 1970.

Scoresby, W. *An Account of the Arctic Regions with a History and Description of the Northern Whale Fishery.* 2 vols. Edinburgh, 1820.

————. "An Inquiry into the Principles and Measures on which Safety in the Navigation of Iron Ships may be reasonably looked for." *BAASR* for 1854.

————. "On the Loss of the *Tayleur* and the Changes in the Action of Compasses in Iron Ships." *BAASR* for 1854.

————. *Journal of a Voyage to Australia and Round the World for Magnetical Research.* Edited by Archibald Smith. London, 1859.

Scoresby-Jackson, R. E. *The Life of William Scoresby.* London, 1861.

Scott, E. *The Life of Captain Matthew Flinders.* Sydney, 1914.

Sharp, Andrew. *Ancient Voyagers in the Pacific.* London, 1957.

Skelton, R. A. *Explorers' Maps.* London, 1958.

Skempton, A. W. *John Smeaton, FRS.* London, 1981.

Smeaton, John. "On some Improvements of the Mariner's Compass." *RSPT*, vol. 46 (1750).

————. "An Account of some Experiments upon a Machine for Measuring the Way of a Ship at Sea." *RSPT*, vol. 48 (1753–54).

Smiles, Samuel. *Lives of the Engineers: Smeaton and Rennie.* London, 1904.

Smith, Archibald. "On the Deviation of the Compass in Wooden and Iron Ships." *BAASR* for 1854.

Smith, Crosbie, and M. N. Wise. *Energy and Empire: A Biographical Study of Lord Kelvin.* Cambridge, 1989.

Spinney, David. *Rodney.* London, 1969.

Stackpole, E. A. *The Sea Hunters.* New York, 1953.

Stamp, Tom, and Cordelia Stamp. *William Scoresby: Arctic Scientist.* Whitby, 1975.

Taylor, E. G. W. "Old Henry Bond and the Longitude." *MM*, vol. 25 (1939).

————. "An Elizabethan Compass Maker." *JIN*, vol. 3 (1950).

————. "Early Charts and the Origin of the Compass Rose." *JIN*, vol. 4 (1951).

————. "The Oldest Mediterranean Pilot." *JIN*, vol. 4 (1951).

————. "Jean Rotz and the Variation of the Compass." *JIN*, vol. 7 (1954).

————. *The Haven Finding Art.* London, 1956.

Thompson, S. P. *Gilbert of Colchester: An Elizabethan Magnetizer.* London, 1891.

————. *Notes on the De Magnete of Dr William Gilbert.* London, 1901.

————. *William Gilbert and Terrestrial Magnetism in the Time of Queen Elizabeth: A Discourse.* London, 1903.

————. *Life of Lord Kelvin.* 2 vols. London, 1910.

————. *The Rose of the Winds: The Origin and Development of the Compass Card.* London, 1913.

Thomson, William. "On Compass Adjustment in Iron Ships, and on a New Sounding Apparatus." *JRUSI*, vol. 22 (1878).

————. "Recent Improvements in the Compass, with Correctors for Iron Ships." *JRUSI*, vol. 24 (1880).

Thrower, N. J. *Maps and Civilization.* Chicago, 1996.

————— (editor). *The Three Voyages of Edmond Halley in the Paramore 1698–1701*. London, 1981 (Hakluyt Society Series 2/156).

Towson, J. T. *Practical Information on the Deviation of the Compass; for the use of Masters and Mates of Iron Ships*. London, 1894.

Toynbee, P., and L. Whibley (editors). *Correspondence of Thomas Gray*. 3 vols. Oxford, 1935.

Waddell, John. "On the Effects of Lightning in Destroying the Polarity of a Mariner's Compass." *RSPT*, vol. 46 (1749–50).

Walker, Ralph. *A Treatise on Magnetism with a Description and Explanation of a Meridional and Azimuth Compass*. London, 1794.

Waters, D. W. "The Lubber's Point." *MM*, vol. 38 (1952).

—————. "Bittacles and Binnacles." *MM*, vol. 41 (1955).

—————. "Early Time and Distance Measurement at Sea." *JIN*, vol. 8 (1955).

—————. *The Art of Navigation in England in Elizabethan and Early Stuart Times*. London, 1958.

Watts, A. J. *The Royal Navy: An Illustrated History*. London, 1994.

Williams, Glyndwr. *The Great South Sea: English Voyages and Encounters 1570–1750*. London, 1997.

Wright, Lawrence. *Clean and Decent*. London, 1966.

Young, Thomas. "Computations for Clearing the Compass of the Regular Effect of a Ship's Permanent Attraction." *Miscellaneous Works of the Late Thomas Young M.D., F.R.S., &c.* Edited by George Peacock. 2 vols. London, 1855.

INDEX

Page numbers in *italics* refer to illustrations.

Aaron Manby, 185–86, 190

Abercorn, Earl of, 101

Account of the Arctic Regions, An (Scoresby), 176–77, 178, 182, 193*n*

Actaeon, HMS, 262*n*

Adams, George, 124–25, 136

Admiralty, British, 22, 26, 51, 81–82, 83, 117, 119, 131, 138, 170, 171–72, 182, 184–85, 219, 228, 253, 256

    Compass Committee of, 196, 206–8, 214, 215, 268–69

    Compass Department of, 218, 226, 230, 257, 270

    compass deviation problem and, 192–93, 195–96

    destroyers developed by, 262–63

    Gray's binnacle and, 231–32, 233

    Knight's compass and, 120–24, 132–33, 150

    liquid compass used by, 268–69

    Napoleon and, 149, 150

    Navy Board and, 78–79, 81, 83, 85, 121, 128, 152

    Paramore voyages and, 78–79, 84–85, 88, 90

    see also Royal Navy

Admiralty Manual for Ascertaining and Applying the Deviations of the Compass Caused by the Iron in a Ship (Evans and Smith), 218

Admiralty Standard Compass, 208–10, 218, 219, 221, 226, 230, 241, 256, 263

Advertisement Necessary to be Observed in the Navigation Up and Down the Channel of England, An (Halley), 95

Africus (southerly winds), 44

Agathermos, 44

agonic line, 62
Airy, George, 196–98, 209, 210, 212, 213, 218, 219, 223, 226, 238
    compass system of, 199–203, 204, 206
    in *Ironside* investigation, 201–3, 205
    in *Rainbow* investigation, 199–201, 206
    Scoresby's feud with, 216–17, 231
Albert, Prince, 226, 228
Alburkali, 191
Alexander II, Tsar of Russia, 251
Alfred the Great, 51
Amalfi, 47, 49, 50–51
annual change, 67*n*
annual winds, 44
Anschütz-Kaempfe, Hermann, 274
Anson, George, 119–21, 124
Apeliotes, 41
Apollo, HMS, 189–90
Ardent, HMS, 263
Argentina, 261
Argyll, Duke of, 248
Aristotle, 43, 46
Armed Neutrality, 146
Arrowsmith, Aaron, 170
Arrowsmith, John, 170
artificial horizon, 122
Association, HMS, 19
astrolabe, 23, 272
Athenaeum, 215, 223, 231
Attempt to Demonstrate That All the Phenomena in Nature May Be Explained by Two Simple Active Principles, Attraction and Repulsion, An (Knight), 107*n*
Audacious, HMS, 245
Augustine of Hippo, Saint, 32
Australia, 131, 149
    Flinders's expedition to, *see* Flinders, Matthew, Australian expedition of
    French expedition to, 149–50, 153–54
    gold rush in, 211–12

Austria, 261
Austro-Hungarian Navy, 270
azimuth compass, 81, 217
    Admiralty Standard Compass as, 208
    function of, 116–17
    magnetic variation and, 64
    Smeaton's design of, 116–17, 124
    Walker's design of, 136–39, 143
Azores, 58, 62

Bacon, Reginald, 257
Bacon, Roger, 32, 39
Banks, Joseph, 138, 144–47
    death of, 175
    Flinders's Australia expedition and, 145–46, 150–51, 165, 166, 170
    Scoresby and, 175, 178–80
Banshee, HMS, 263
Barlow, Peter, 184–85, 192
Barlowe, William, 14, 63–64, 67, 265–66
Barlow plate, 184, 192
Barracuda Tanker Corporation, 28
Barrow, John, 179
Bass, George, 143–44, 152
Baudin, Nicolas, 149, 154, 155
Beaufort, Francis, 193–95, 196, 207, 210, 268
Beechey, F. W., 226
Benbow, John, 80, 82, 87
Bermuda, 92
Bertelli, Timoteo, 51
binnacles, 100
    double, 230
    Gray's, 231–32, 233
    Thomson's, 241–42, *242*
Biographia Britannica, 75
Bird, John, 111
Blackbarrow, Peter, 68
Black Death, 51
    of 1665, 70
Black Sea, 44, 227
Blandy family, 239
Bligh, William, 133, 137, 140*n*, 142–43, 145, 152
Board of Investigation, 29

Board of Longitude, 20, 120, 124, 136, 138, 140, 182
Board of Trade, 213, 226, 232–33
Bodleian Library, 38, 56
Bodley, Thomas, 56
Bond, Henry, 68
Book of Bearings (Flinders), 153, 157–60, 162, 163, 166
Boreas, 41, 44
Borough, Stephen, 57, 59
Borough, William, 57, 58, 59–60, 61, 62, 64, 67
Bounty, HMS, 133, 137, 140*n*
Bourne, William, 24
bowline, 46
Boxer, HMS, 263
"boxing the compass," 54
Brande's Quarterly Journal, 182
Brazil, 261
Bridgewater, 162
Britannic, 250
British Association for the Advancement of Science, 194, 213, 215, 222
British India Steam Navigation Company, 244–45, 250
British Museum, 56
Knight as Director of, 131–32
British Petroleum, 28
Bruiser, HMS, 263
Brunel, Isambard Kingdom, 228–29, 243
Burchett, Josiah, 84

Cabot, Sebastian, 62
Caesar, Julius, 70
Cairns, Nathan, 201
Campbell, Alexander, 121, 123
Canton, John, 105
Cape of Good Hope, 90, 94, 164
Cape Verde Islands, 82, 88
Captain, HMS, 249, 253
Castle Packets, 245
Castro, João de, 139
Catalogue of the Southern Stars (Halley), 75
Cato, 162–63
Cavendish, Thomas, 265

"Celestial Bed," 106–7
Centurion, HMS, 119–20
Champlain, Samuel de, 24
Chancellor, Richard, 56, 57
Chanda, 250
"Channel Course," 96
Chao Ju-Kua, 13
Charger, HMS, 263
Charles II, King of England, 72, 73, 75
charts, 63
circumpolar, 57
color in, 48, 58
Dee's design of, 57
Halleyan Lines in, 92–95, *93, 94*
magnetic variation first shown in, 57–58
of Mediterranean Sea, 48–49
symbols used in, 22
wind rose of, 48–49, 50, 52, *52,* 54
winds indicated in, 48–49
Chaucer, Geoffrey, 52
Chetwynd, Louis W. P., 257–58, 270–71
China, 36–38, 150, 204, 261
Christian, Fletcher, 140*n*
Christie, Samuel Hunter, 192, 207, 215
chronometer, 209–10
Chu Yü, 37
circumpolar chart, 57
City of Baltimore, 222
City of Dublin Steam Packet Company, 191–92
Clark, James, 74
clipper ships, 47
Cocking, Mr., 206–7
Colossus, HMS, 245
compass adjuster, profession of, 231, 232
compass cards, 25, 50, *53*
Dee's design of, 57
dipping and, 60–61
inaccuracies in, 63–64
in liquid compasses, 266, 269–71
swirl problem and, 270
in Thompson compass, 239, *240,* 241
360–degree, 54

Compass Committee, Admiralty, 196, 214, 215, 268–69
  first meeting of, 206–8
Compass Department, Admiralty, 210, 218, 226, 230, 257, 270
compasses:
  in China, 37–38
  direction points of, 52–53, *53*
  first patented, 126–27
  as gyrocompass backup, 275–76
  in high latitudes, 176, 180
  invention of, 37–38, 51
  markings of, 52–53
  in submarines, 274
  true vs. magnetic north in, 54
  see also charts; deviation, compass; dip, magnetic; variation, magnetic; *specific types of compasses*
compass needles, 31, 34–39, 58, 59, 63, 64, 65–66, *66*
  Chinese adaptation of, 36–38
  dipping and, 67, 68
  of *Dover,* 112–13
  in liquid compasses, 266–68
  multiple, 208–9, *219*
  Neckham's description of, 34, 50
  rectangular, 113–14, 116
compass rose, 54
"Concerning the Differences in the magnetic Needle on board the *Investigator,* arising from an Alteration in the Direction of the Ship's Head" (Flinders), 166
Conflict, HMS, 263
Conqueror, HMS, 245
contour lines, 93
Cook, James, 95, 128, *129,* 139–40, 141, 143, 145, 152, 194
  Flinders compared with, 142
  Knight's compass tested by, 127–28, 130–31
Creak, Ettrick W., 257, 271, 275
Creuze, Augustin, 205
Crimean War, 227, 230
cross staves, 23

Crow, Francis, 266–68, 270
Crusades, 44, 47
Crystal Palace, 225
Cumberland, 163–65
Cunard line, 245
Curve Lines (Halleyan Lines), 92–95, *93, 94*

D'Arifat, Delphine, 168–69
D'Arifat, Madame, 168
Daring, HMS, 263
Dasher, HMS, 263
Davis, John, 57, 74
"dead beat" escapement, 111
dead-reckoning:
  dry compass in, 25–26
  lead line in, 21–24
  log in, 24–25
dead-reckoning position (D.R.), 21
Decaen, Charles, 165–66, 168, 169–70
Dee, John, 55–56
deep-sea lead line, 22–23
De Magnete (Gilbert), 65, *66*
Demosthenes, 43
De Naturis Rerum (Neckham), 33–34
Denmark, 146, 261
Dent, E. J., 264
depth-sounder, 21, 28
Desire, 265
Desperate, HMS, 263
destroyers, 262–63
De Utensilibus (Neckham), 33–34
Deutschland, 252, 274
Devastation, HMS, 245
deviation, compass:
  Admiralty's investigation of, 192–93, 195–96
  Airy's *Rainbow* investigation and, 199–201
  Barlow's plate and, 184
  disasters associated with, 187–90
  Flinders's bar idea and, 169, 172–73
  Flinders's investigation of, 152–53, 169, 172–73
  Flinders's paper on, 166, 169

heeling error and, 206, 208, 278
in high latitudes, 180–81
Ironside and, 201–3, 205
octantal, 279
proximity of iron and, 138–40
quadrantal, 206, 279
Scoresby investigation into, 178,
    180–81, 213–16, 219–22
semicircular, 206, 277–78
sextantal, 206, 279
Young's theory of, 182, 183–84
deviation card, *222*
Deviation Table, 210, 230–31
Dickens, Charles, 226
Dictionary of National Biography, 29, 67,
    68
dip, magnetic, 74–75, 172
    compass card and, 60–61
    compass needle and, 67, 68
    in Flinders's Australia expedition,
        152–53
    in high latitudes, 180–81
    soft iron and, 181
    in Young's theory, 183–84
Discourse of the Variation, A (Borough), 61
diurnal change, 67*n*
Dodson, James, 126
Dolphin, HMS, 127–28
dome-top compass, 271
double binnacle, 230
Dover, 99–101, 107, 109–10, 112–13, 115
Downie, Murdo, 138–39
Dragon, HMS, 263
Drake, Francis, 62, 265
Dreadnought, HMS, 254, 257*n*
dry compass, 25–26, 37, 50, 100, 208,
    258–59, 274
Dryden, John, 65
Ducie, Earl, 249*n*
Dufferin, Lord, 249

Eagle, HMS, 19
earthkins, 65–66
East India Company, 73, 150
    Secret Committee of, 203

east variation, 27
Eddystone Lighthouse, 124, 130
Egypt, 43, 44, 49
Elements of Navigation, The (Robertson),
    126, 251
Elizabeth I, Queen of England, 55–56,
    65, 265
Encyclopaedia Britannica, 165
Endeavour, 128, 130–31, 141, 145, 147
Erebus, HMS, 207
Esk, 178, 180
Etesian winds, 44
Euros, 41
Evans, Frederick J. O., 218, 231, 239, 279
Evelyn, John, 66, 80
Exiles' Line (Kipling), 244*n*
Eyre, Edward John, 151–52

Fairbairn, William, 188–89
Falconbird, 87–88
Falconer, William, 131
Farrer, Thomas, 232–33
Ferren, Bran, 16
Ferret, HMS, 263
Fervent, HMS, 263
Field, Mostyn, 257
Fighting Temeraire, The (Turner), 226
Firebrand, HMS, 19
First Crusade, 47
Fisher, John H. "Jacky," 228, 254–55, 256
    destroyers development and, 262–63
"Fishpond," 257
Flamsteed, John, 72–73, 96–97
Flinders, Ann, 170, 172
Flinders, Matthew, 133, 138, 140,
    141–47, 151, 180, 181, 184,
    186, 210, 218, 279
    Australia expedition of, *see* Flinders,
        Matthew, Australia expedition
        of
    background of, 142
    bar of, *see* Flinders bar
    Bass's friendship with, 143–44
    compass deviation investigated by,
        152–53, 169, 172–73

Flinders, Matthew (*continued*)
  Cook compared with, 142
  death of, 172
  Mauritius captivity of, 164–70
  in *Providence* voyage, 142–43
  shipwrecked, 162–64
Flinders, Matthew, Australian expedition
  of, 141–70
  Banks and, 145–46, 150–51, 165,
    166, 170
  Book of Bearings in, 153, 157–60,
    162, 163, 166
  Flinders's Mauritania captivity in,
    164–70
  French expedition and, 149–50,
    153–54, 161
  Géographe encountered in, 153–54
  Great Barrier Reef in, 158–59
  idea for, 143–44
  magnetic dip anomaly in, 152–53
  onset of, 141–42
  proposal for, 145–46
  shipwreck in, 162–64
  Sydney stopover in, 154–57
  Timor leg of, 159–60
  unseaworthiness of *Investigator* in,
    151, 158–59, 161
Flinders bar, 169, 172–73, 182, 218, *242,*
  243, 279
fluxgate (inductor) compass, 15, 275–76
Foam, 249*n*
Folger, Mayhew, 133
Folkes, Martin, 103, 120
Foreign Office, British, 150
Fortune, HMS, 121
Fothergill, John, 105
Foudroyant, HMS, 116
France, 20, 58, 62, 146, 227, 245
  ironclads developed by, 227, 229–30
  Mauritius as colony of, 166–68
  New Holland expedition of,
    149–50, 153–54
  torpedo boat development and,
    261–63
Francis, 144

Franklin, Benjamin, 107
French Academy, 111
Frobisher, Martin, 57

Galileo Galilei, 65
Garland, 268
Garryowen, 191–93, 197
Gellibrand, Henry, 67–68, 78
General Steam Navigation Company,
  196, 200–201
Genoa, 47, 49, 50
Géographe, 149–50, 153–54, 155, 156
George III, King of England, 110, 145
George VI, King of England, 275
Germanic, 250
Germany, 245, 261
Gibbons, Edward, 102*n*
Gilbert, Humphrey, 57
Gilbert, William, 64–67, *66,* 101
  earthkins of, 65–66
Gioia, Flavio, 51
Glasgow News, 239
Gloire, 230
Glory, HMS, 121, 138–39
Gooch, Daniel, 228
Good Words, 238
Graham, George, 111–12
Graham, James, 106–7
Gray, John, 231–32
Gray, Thomas, 132
Gray's binnacle, 231–32, 233
Great Australian Bight, 151–52
Great Barrier Reef, 158–59, 164–65
Great Britain, 20, 51, 58, 62, 146,
    149–50, 227, 231
  Pax Britannica of, 243–44, 245
  torpedo boat development and,
    261–62
  see also Admiralty, British; London;
    Royal Navy
Great Britain, 228–29
Great Embassy, 79
Great Exhibition of 1851, 225–27
Great Fire of London, 70
Greco (northeast wind), 46

Greece, 261
Greenland, 58, 178–79
Greenwich Meridian, 72
Grenville, 128
Gwynn, Nell, 72*n*
gyrocompass, 28
  magnetic compass and, 273–76

Hadley, John, 20
Hakluyt, Richard, 57
Hall, Christopher, 57
Halley, Edmond, 27, 68, 69–98
  background of, 70–71
  Curve Lines of, 92–95
  death of, 97
  in English Channel survey, 95–96
  longitude-finding method of, 71–72
  at Mint, 79
  at Oxford, 73
  in *Paramore* expeditions, *see Paramore,*
    HMS, voyages of
  personality of, 75
  Peter I and, 80
  as polymath, 69–70, 71
  as royal astronomer, 96–97
  in St. Helena expedition, 73–75
  World Chart of, 94–95, 98, 136
Halleyan Lines (Curve Lines), 92–95, *93,*
  94
Hamelin, Jacques, 155, 156–57
Hamilton, John, 9
Hanover, Electress of, 79
Harding, Fisher, 79
hard iron, 184, 185–86, 193, 278
Hardwick, Mr., 91
Hargraves, Edward, 211
Harrington, 154*n*
Harrison, Edward, 83–84
Harrison, John, 20, 111
Hasty, HMS, 263
Hatfield House, 57–58
Hatton, Christopher, 55*n*
Havoc, HMS, 263
Hawkins, John, 57
Hawksmoor, Nicholas, 199

heeling error, 206, 218, 278
Helmholtz, Hermann von, 248–49
Henry Hughes & Son, 252*n*
Hercules, HMS, 245
Homer, 44
Hooke, Robert, 72
Hooper, 239
Hope, 162–63
Hornet, HMS, 263
Hotspur, HMS, 245
Howe, Richard, 121
Hugo, Victor, 14
Hunter, HMS, 263
Hunter, John, 143
Hurd, Thomas, 170
Hutchinson, William, 122–23, 126
Huxley, Thomas, 67
Hydrographic Office, 153

Iceland, 61–62
Idea Longitudinis (Harrison), 83
Île de France, *see* Mauritius
India, 243–44, 250
induced magnetism, 180, 206, 218, 279
inductor (fluxgate) compass, 15, 275–76
Inflexible, HMS, 245
Ingenhousz, John, 265–66
International Exhibition of 1862, 230
Investigator, HMS, 141–42, 147, 149, 150,
    153, 155, 157, 165, 166, 168
  unseaworthiness of, 151, 158–59,
    161
Invincible, HMS (18th-century warship),
    138*n*
Invincible, HMS (19th-century warship),
    245
Ireland, 58
iron:
    hard, 184, 185–86, 193, 278
    soft, 181, 184, 185–86, 193, 218–19,
      278, 279
ironclad warships, 227, 229–30
Ironside, 201–2, 278
isogones, 93
Italy, 261, 270

Jameson, Robert, 176–77
Japan, 37–38, 261
Jarvis, Thomas Best, 207
Jenkin, Fleeming, 237
Jermy, John, 121
Jerusalem, Latin Kingdom of, 47–48
John Randolph, 190
Johnson, Edward, 192–94, 196, 197, 207, 210, 226
Johnson, Samuel, 104, 112
Jones, Inigo, 199
Jonson, Ben, 66
Joule, James P., 239
Journal of a Voyage to Australia and Round the World for Magnetical Research (Scoresby), 223
Jumper, William, 21, 25

Kaikos, 41
Kelvin Hughes, 252*n*
Kelvin of Largs, Baron, *see* Thomson, William
Keroualle, Louise de, 72
King, Philip Gidley, 155, 157–58, 161, 163
Kipling, Rudyard, 244*n*
Knight, Gowin, 98, 107, 215, 238, 251, 256–57
    Anson's visit with, 120–21
    artificial magnet of, 102–6, *106*
    background of, 102–3
    at British Museum, 131–32
    Cook's meeting with, 128–30
    death of, 132
    Dover's compass inspected by, 107, 109–10, 112–13
    magnetic compass of, 114–15, 122–28, *125,* 130–33, 138, 208
    rectangular compass needle of, 113–16
    Royal Society and, 102–4
Kyrenia II, 43

Lady Nelson, HMS, 155, 157–59
Laird, John, 190–94, 196, 203

Lalla Rookh, 235–39, *237,* 247, 248, 253
Lane, Frederic, 50
Latin Kingdom of Jerusalem, 47–48
latitude, 67
lead line, 21–23, 24
    markings on, 22–23
Lecky, Squire Thornton Stratford, 14, 250–51, 254
leeway, 21
Leicester, Earl of, 55*n*
Lennox, HMS, 21
Levant compass, 63
Levante, 46
Lever, Darcy, 53, *125*
lever escapement, 112
Libeccio, 46
Lightning, HMS, 258–59
Lips, 41
liquid compass, 258–59
    Chetwynd's design of, 258, 270–71
    compass card of, 266, 269–71
    compass needle of, 266–68
    Crow's design of, 266–68, 270
    development and evolution of, 265–72, *267*
    dry compass as backup for, 275–76
    dry compass supplanted by, 258–59, 263–64
    freezing problem and, 266
    Ritchie's design of, 258, 269
    Royal Navy and, 265–66, 268–69
    in sea trials, 263–64
    swirl problem and, 270–71
    Thomson compass and, 257–58, 263–64
    of today, 271–72
    torpedo boat development and, 263–64
Livadia, 251
Liverpool Compass Committee, 217–18, 243
local attraction (local magnetic anomaly), 67*n*
lodestones, 26, 31–36, 56–57, 58, 101
    artificial, *see* compass needles

Bacon's experiments with, 32
characteristics of, 31
commercialization of, 101–2
first record of, 33
in medieval lore, 32–33
Neckham's description of, 33–34
as perpetual clock, 38–39
log, 24–25, 122
London, 59–60, 110–11
Great Exhibition of 1851 in, 225–27
Great Fire of, 70
magnetic variation at, 59–60, 71
London Chronicle, 170
longitude, 20, 67, 83, 136, 138
lunar-distance method of finding, 71–72
magnetic variation and, 62, 67–68
saronic cycle and, 97
Longitude and Latitude discovered by the Inclinatory or Dipping Needle (Whiston), 68
Longitude Found, The (Bond), 68
Longitude Not Found, The (Blackbarrow), 68
loran, 28
Lord Dundas, 187–89
Louis XIV, King of France, 79
Louise, Princess, 248
Lous, Christian, 209
loxodromes (rhumb lines), 49
lubber's line, 100
lunar-distance method, 71–72
Lynx, HMS (destroyer), 263
Lynx, HMS (18th-century warship), 138*n*

Maestro (northwest wind), 46
Magellan, Ferdinand, 62
Magnaghi, G. B., 270
Magnet Creek, Ark., 31
Magneticall Advertisments (Barlowe), 63
magnetic dip, *see* dip, magnetic
magnetic induction, 180, 206, 218, 279
Magnetic Lady, The (Jonson), 66
Magnetic Observatory, 197, 200

magnetic variation, *see* variation, magnetic
"Magnetism of Ships" (Flinders), 172
magnetite, *see* lodestones
Magnus, Olaus, 14
manuscript rutter, 23
Margaret, 155
Maricourt, Pierre de, *see* Petrus Peregrinus
Marine Board of Liverpool, 213
marine chronometer, 20
"marine diver," 177
Mariners' Magazine, 139
Mariner's Mirrour (Wagenaer), 23–24
Martin, William, 144
Mary II, Queen of England, 78
Maskelyne, Nevil, 138
Maurice of Nassau, Prince, 167
Mauritius (Île de France), 150, 167
Flinders's captivity in, 164–66, 168–69
May, W. E., 96
Mediterranean Sea, 42–46
charts of, 48–49
sailing season of, 42–43, 49
winds of, 41–42, 44, *45,* 46
Mêng Chhi Pi Than (Shen Kua), 36
Mercator, Gerardus, 56
mercurial pendulum, 111
meridional compass, 57
Mezzodi (south wind), 46
Middleton, Benjamin, 78
Monitor, USS, 269
Morris, William, 226
Mountaine, William, 126
Mudge, Thomas, 112
Muscovy Company, 56, 57, 58

Napoleon I, Emperor of France, 146, 150
Naturaliste, 149–50, 154, 155, 156
Nautical Almanac, 182
Naval Chronicle, 142, 172
Navigators Supply, The (Barlowe), 63
Navy, U. S., 26*n,* 209, 218, 258, 269

Navy Board, British, 78–79, 81, 83, 85, 121, 128, 152
Neckham, Alexander, 13, 33–34, 39, 271
    compass needle described by, 34, 50
Nelson, Horatio, 146, 194, 226
Nemesis, 203–6, 228
Nepean, Evans, 151
Netherlands, 261
Newe Attractive (Norman), 61, 65
Newfoundland, 61–62, 128, *129*
New Holland, *see* Australia
Newton, Isaac, 71, 74, 96, 101
New World, 61–62
New Zealand, 131
Nigeria, 145
Norman, Robert, 60–64, 65, 67
Northampton, HMS, 255
Norway, 58, 261
Notos, 41
Nullarbor Plain, 151

octantal deviation, 279
"On the Anomaly in the Variation in the Magnetic Needle, as Observed on Ship-Board" (Scoresby), 178
"On the Loss of the *Tayleur,* and the changes in the Compass in Iron Ships" (Scoresby), 215–16
Opium War, 204*n,* 228
Ostro, 46
overhead (telltale) compass, 202, 203
Oxford University, 75
oxide of iron, *see* lodestones

Paixhans, Henri-Joseph, 227
Paramore, HMS, 77, 79, 87, 90, 92
    in English Channel survey, 95–96
    sailing qualities of, 80–81
Paramore, HMS, voyages of:
    Admiralty's instructions for, 81
    Halley's conflict with officers in, 82–83, 84
    Halley's navigation in, 89–90
    icebergs encountered in, 89–90

magnetic variation investigations in, 83, 85, 92–93
    objectives of, 78, 81
    Peter I and, 80
    proposal for, 78
    White's death in, 88
Park, John, 163
Park, Mungo, 145
Pax Britannica, 243–44, 245
Paxton, Joseph, 225
Peace of Amiens, 164
Peichel, Joseph Von, 270
Peninsula and Oriental Steam Navigation Company (P&O), 244, 250
Pepys, Samuel, 66
Peter I, Tsar of Russia, 79–80
Petrus Peregrinus (Pierre de Maricourt), 38–39
Philadelphia Centennial Exhibition, 258
Phillip, Arthur, 156
Philosophical Transactions (Royal Society), 97–98, 102, 111–12, 117, 166, 194
Phingchow Table-Talk (*Phing-Chou Kho Than*) (Chu Yü), 37
Phoenix, HMS, 21
pilot books, 49
Pisa, 47, 49, 50–51
Pitcairn Island, 133
plague (Black Death), 51
    of 1665, 70
planisphere, 62
Plato, 43
Pliny the Elder, 44, 70
pocket sundials, 58–59
Poisson, Siméon-Denis, 218
polarity, 31
Pollard Rock, 29
Ponente, 46
Porpoise, HMS, 155, 161–62, 164
portolans, 49
Portugal, 58, 62, 189
Prince Frederick, 127
Providence, HMS, 137–38, 140, 142–43
Provins, Guyot de, 35

Prussia, 146
Punch, 225

quadrant, 23, 272
quadrantal deviation, 206, 279
Queen, HMS, 138*n*

radar, 23, 28
radio-direction finder, 28
Rain, Steam and Speed (Turner), 226–27
Rainbow, HMS (warship), 121
Rainbow (steamship), 196, 197–201, 206
Raleigh, Walter, 57
Ramsden, Jesse, 112
reflecting quadrant, 20
Regiment for the Sea, A (Bourne), 24
Reliance, HMS, 143, 145, 149
Reliance (East Indiaman), 190
Rennell Current, 26–28
Rennie, George, 187–88
Resolution (Cook's vessel), 139, 140*n*
Resolution (whaling ship), 176
rhumb lines (loxodromes), 49
Richard I (the Lion-Hearted), King of
    England, 33
Ritchie, E. S., *267,* 269
Ritchie's Floating Compass, 258
Robertson, George, 127–28
Robertson, John, 126, 251
Robur Carolina (constellation), 75
Rocket, HMS, 263
Rodney, George, 121, 123
Romantic Movement, 142
Rome, Imperial, 43, 44
Romney, HMS, 19
rose of the wind, *see* wind rose
Ross, James Clark, 207, 209, 214,
    215
Ross, John, 183, 277
Royal African Company, 69, 87, 91
Royal Astronomical Society, 195
Royal Charles, HMS, 71
Royal Charter, 219–21, 223
Royal Geographical Society, 195
Royal Institution, 194

Royal Navy, 19–20, 75, 113, 138, 146,
    152, 180, 191, 194, 197
    Arctic expedition of, 179–80, 183
    destroyers developed by, 262–63
    gyrocompass adopted by, 274
    ironclads in, 227–30
    liquid compass used by, 265–66,
        268–69
    Paramore voyages and, 77–79
    standard compass of, 208–10
    Thomson compass and, 252–54,
        257–58
    see also Admiralty, British
Royal Observatory at Greenwich,
    72–73, 96–97, 111, 195
    Magnetic Observatory of, 197, 200
Royal Society, 70, 75, 96, 101, 115–16,
    117, 138, 166, 178, 193, 194,
    195, 231, 232–33, 238, 239
    Knight and, 102–4, 115
    Paramore expeditions and, 78, 83, 92,
        97–98
Royal United Service Institute, 241, 242
Royal Yacht Squadron, 1, 249*n*
rudder, 46–47
Rundell, W. W., 217–18
running the latitude down, 90
Russia, 79, 146, 227, 245, 261
Russia, 240, 250

Sabine, Edward, 207, 209
sailing, 42–47
    charts in, 48–49
    and development of tramp steamer,
        245
    evolution of, 46–47
    season for, 42–43, 46, 49–50
    steering with compass in, 50
sailing directions, 49
St. George, HMS, 19
St. Helena, 73, 90
Sallymen, 82, 87–88
sandglass, 23
Sandwich, Lord, 66
saronic cycle, 97

Savery, Servington, 102
Schanck, John, 157
Science Museum, London, 101
Scilly Islands, 19, 20–21, 26, 28–29, 95, 96
Scoresby, William, 175, 186, 193*n,* 215, *219,* 279
    Airy's feud with, 216–17, 231
    background of, 175–76
    Banks and, 175, 178–80
    Barrow's meeting with, 179
    compass deviation investigated by, 178–81, 213–16, 219–22
    death of, 223
    Jameson and, 176–77
    "marine diver" invention of, 177
    in *Royal Charter* voyage, 219–22
    Royal Navy's Arctic expedition and, 179–80
    ships' magnetic signatures noted by, 215–17
Scotland, 58
Sea Grammar (Smith), 139
Searchthrift, 59
Secret Committee of the Honourable East India Company, 203
secular changes, 27, 67–68, 78
    Halleyan Lines and, 94, *94*
semicircular deviation, 206, 277–78
Seven Stone rocks, 28
sextant, 28, 112
sextantal deviation, 206, 279
Shark, HMS, 263
Sharp, Andrew, 14
Shen Kua, 36
Shovell, Cloudesley, 19–21, 25, 27, 28, 29, 68, 82, 83, 96
Sirocco, 46
Skiron, 41, 42
Sloane, Hans, 131–32
Smeaton, John, 114–16, 122, 130
Smith, Archibald, 209, 218–19, *219,* 223, 233, 235–38, 242, 279
Smith, John, 139
Smith, W. H., 252

Socrates, 43
soft iron, 181, 184, 185–86, 193, 218–19, 278, 279
South Africa, 245
Spain, 62, 119, 120
Spanish Armada, 265
Sparrowhawk, HMS, 263
Speedy, 155
Spencer, Earl, 146–47
Sperry, Elmer, 274
Spieghel der Zeevaerdt (Wagenaer), 23–24
Spitfire, HMS, 263
Stephenson, George, 243
Sturmy, Samuel, 139
submarines, 274
Suez Canal, 243
sundials:
    pocket, 58–59
    universal, 137
super magnetism, 31
Supply, 155
Surcouf, Robert, 167–68
Swallow, HMS, 127
Swan, HMS, 121
Sweden, 146, 261
Swordfish, HMS, 263
Symonds, William, 229

"Table of Corrections for clearing the Compass of the regular effect of a Ship's Permanent Attraction" (Young), 183
Tabula de Amalpha, 50
Tahiti, 128, 130–31, 140*n,* 142
Tayleur, 212–13, 215–16, 223
telltale (overhead) compass, 202, 203
terrellas, 65–66, 101
Terror, HMS, 207
Thetis, HMS, 190
Thistle, John, 152
Thomson, William, 219, 235–43, 245, 263, 264
    binnacle designed by, 241–42, *242,* 249–52, 258
    in *Captain* inquiry, 249, 253

compass card of, 239, *240,* 241
compasses designed by, 238, 239–43
Fisher as advisor to, 256
marketing campaign of, 249–52,
    255–57
marriage of, 239–40
Smith and, 235–38
Thomson binnacle, 241–42, *242,*
    249–52, 258
Thomson compass:
    card of, 239, *240,* 241
    development of, 238, 239–43
    liquid compass and, 257–58, 263–64
    marketing of, 249–52, 255–57
    Royal Navy and, 252–54, 257–58
    testing of, 247–49
    Times on, 255–56
Thrasher, HMS, 263
Thunderer, HMS, 245
Times (London), 170, 204*n,* 206, 215,
    216, 232
    on Thomson compass, 255–56
Timor, 159, 164
Tom Thumb, 144
Topaz, 133
torpedo boats, 261–63
Torpedoes & Torpedo Vessels (Royal Navy),
    264
Torrey Canyon, 28–29
Tower of Flies, 45
Tower of the Winds, 41, 44, 46
Tramontana, 46, 52
tramp steamer, 245
Traveler's Jewel, 64
"Treatise on the Astrolabe" (Chaucer), 52
Tristan da Cunha, 89–90
Turkey, 227, 261
Turner, J. M. W., 226–27
Tuttell, Thomas, 102

Union Oil of California, 28
Union Steamship, 245
United Services Journal, 205
United States, 261
    shipbuilding in, 250

United States Navy, 26*n,* 209, 218, 258,
    269
Unity, 74
Universal Dictionary of the Marine, An (Fal-
    coner), 131
universal sundial, 137
Urban II, Pope, 47

Vanbrugh, John, 199
Vanguard, HMS, 275
variation, magnetic, 27–28, 54, 78, 90
    abnormal, 67–68
    agonic line and, 62
    azimuth compass and, 64
    discovery of New World and, 61–62
    English Channel survey and, 95–96
    finding longitude and, 62, 67–68
    first charting of, 57–58
    Frobisher's voyage and, 57–58
    Halleyan Lines and, 92–95, *94*
    in high latitudes, 176, 180
    irregularity of, 62–63, 67
    at London, 59–60, 71
    measuring of, 58–59, 74–75
    in modern charts, 63
    offsetting of, 58–59
    Paramore voyage investigation of, 83,
        85, 92–93
    pocket sundial and, 58–59
    proximity of iron and, 138–40,
        152
    secular changes and, 67–68
variation compass, 60
Varley, Cromwell, 237
Venice, 47, 49, 50
Vespucci, Amerigo, 49
Victoria, 272
Victoria, Queen of England, 226, 236*n,*
    243, 248
Victory, HMS, 133, 254
Virago, HMS, 263
Virginia, CSA, 269*n*
Vitry, Jacques de, 34–35, 44–45
*Voyages de la Nouvelle France Occidentale,
    Les* (Champlain), 24

Voyage to Terra Australis, A (Flinders),
     172–73
Vulture, HMS (destroyer), 263
Vulture, HMS (sloop), 121

Waddell, John, 100–101, 109
Wagenaer, Lucas, 23
Waldseemüller, Martin, 49
Wales, William, 140, 144
Walker, Ralph, 136–39, 143, 171
Warrior, HMS, 230–31
west variation, 27
Whiston, William, 68, 136
White, James, 249, 251–52
White, Manley, 88
Whitehead self-propelled torpedo, 259
White Star line, 245, 250
Wiles, James, 138
wind rose, 48–49, 50, 52, *52,* 54
winds, 41–45, *45,* 51–52

depicted in charts, 48–49
personalized names of, 41–42, 46
Wood, Anthony, 71
World Chart of Magnetic Variation,
     94–95, 98, 136
Wren, Christopher, 72, 101, 199
Wright, Gabriel, 266
Wrinkles in Practical Navigation (Lecky),
     251
Wu Tzu-Mu, 37–38
Wyatt, Thomas, 121

Xenophon, HMS, 146–47

Yarrow, Alfred, 262, 263*n*
Young, Thomas, 182–84, 186
Young Sea Officer's Sheet Anchor, The
     (Lever), 53–54, *125*

Zephyros, 41–42